BEAUTIFUL
GARDENS

written by
Eric A. Johnson & Scott Millard

photography by
Scott Millard

illustrations by
Don Fox

IRONWOOD PRESS

Book Design & Production
Millard Publishing Services,
Tucson, Arizona

Additional Photography
Tom Craig, page 88
Francis Ching, page 57
Bill Dewey, page 55
Eric Johnson, pages 11, 77 (bottom),
79, 80, 90
The J. Paul Getty Museum, page 45
Christine Keith, pages 18, 23, 28
Charles Mann, page 92
Peter Rogers, page 75
University of Nevada, Las Vegas,
page 91

Printing 10 9 8 7 6 5 4 3 2 1

ISBN 0-9628236-0-0

Library of Congress Catalog Card
Number 90-085370

Cover and title page: The Hortense
Miller Garden in Laguna Beach,
California.

Contents page: The Lummis
Garden, Los Angeles.

The information in this book is true
and accurate to the best of our
knowledge. It is offered without
guarantees on the part of the
authors, who disclaim liability in
connection with the use of this
information.

Address inquiries to:
IRONWOOD PRESS
2968 West Ina Road #285
Tucson, Arizona 85741

Dedication
This book is thoughtfully dedicated to the
memory of Walter L. Doty, who taught us
how to walk in the reader's shoes.

A sincere thank you to all who
volunteered information and insight.
A special thanks to those who told
us of an unknown garden.

ARIZONA

Wayne A. Hite, Executive Director,
The Arboretum at Flagstaff,
Flagstaff

Ray Kimball, Director, Visitors
Center, and Eva Watson Quist,
Temple Gardens Consultant,
Arizona Temple Gardens, Mesa

Robert E. Breunig, Ph.D., Executive
Director, Desert Botanical Garden,
Phoenix

Volunteers of Valley Garden Center,
Phoenix

Steve Whitley, Sharlot Hall
Museum, Prescott

Carol Crosswhite, Curator, Boyce
Thompson Southwest Arboretum,
Superior

Luray Yule and Mark Dimmitt,
Arizona Sonora Desert Museum,
Tucson

Laura Hnilo, Sabino Canyon, Tucson

Alice Cartey Herman, Executive
Director, Tohono Chul Park, Tucson

Marty Eberhardt, Executive
Director, Tucson Botanical Gardens,
Tucson

Megan Reid, Director, Rio Colorado
Division of Arizona Historical
Society, Century House, Yuma

CALIFORNIA

Staff of Los Angeles State &
County Arboretum, Arcadia, South
Coast Botanical Gardens, Palos
Verdes Estates, and Virginia
Robinson Gardens, Bel Air

Mark C. Jorgenson, naturalist, Anza
Borrego, Borrego Springs

Rebecca Coughman, Public
Relations Director, Rancho Santa
Ana, Claremont

Wade Roberts, Garden Director,
Sherman Botanical Gardens, Corona
del Mar

Bonnie Taylor Smith, Marketing
Manager, Rogers Gardens, Corona
del Mar

Gilbert A. Voss and Diane Goforth,
Quail Botanical Gardens, Encinitas

Celia Kutcher, Curator-Taxonimist,
Fullerton Arboretum, Fullerton

Charles J. O'Neill, Museum
Scientist, Irvine Botanical Gardens,
Irvine

Mary Ann Arnett, President,
Descanso Gardens, La Canada

Andrea Stoner and Fred Lang,
Landscape Architect, Hortense
Miller Gardens, Laguna Beach

John A. Crossman, Chief Ranger
High Desert District, Antelope
Valley Poppy Preserve, Lancaster

Joe McCumminis, Interpretive
Ranger, La Purisima, Lompoc

Pamela Seager, Executive Director,
Rancho Los Alamitos, Long Beach

Ellen Calomiris, Executive Director,
Rancho Los Cerritos Long Beach

Exposition Park, Los Angeles

David Verity, Hannah Carter
Japanese Garden and Mildred
Mathias Garden, Los Angeles

Carol Doujherty, Lummis House,
Los Angeles

Andrea M. Leonard, J. Paul Getty
Museum, Malibu

Lou Feiring, Public Relations
Director, The Living Desert, Palm
Desert

Ron Baetz, Desert Water Agency,
Palm Springs

Patricia Moorten, Director, Moorten
Botanical Garden, Palm Springs

Tom Ash, Director, Landscapes
Southern California Style, Riverside

J.G. Wainer, Director, UCR
Botanical Garden, Riverside

Kathy Kalas Puplava, horticulturist,
Balboa Park, San Diego

Bette Gorton, Naiman Tech Center
Japanese Garden, San Diego

Charles A. Coburn, horticulturist,
San Diego Zoological Garden, San
Diego

James Folsom, Curator, The
Huntington Library, San Marino

David B. Donahue, Hearst's Castle,
San Simeon

Grant Castleberry, Landscape
Architect, Alice Keck Gardens,
Santa Barbara

Anne Steiner, Santa Barbara Botanic
Garden, Santa Barbara

Mark F. Hoefs, Garden Director,
Wrigley Botanical Gardens, Santa
Catalina Island

Janice Busco and Melani Baer,
horticulturists, Theodore Payne
Foundation, Sun Valley

William Truesdell, Chief Naturalist,
Joshua Tree National Monument,
Twenty-Nine Palms

Desie L. Maze, Rose Hills, Whittier

NEVADA

Tracy Little, Manager, Ethel M
Botanical Garden, Henderson

Peter Duncombe, Desert
Demonstration Garden, Las Vegas

Dennis Swartzell, Superintendent,
UNLV Arboretum, Las Vegas

NEW MEXICO

Lynda McBride, Director,
Albuquerque Garden Center

Jane Mygatt, Manager, UNM
Biology Greenhouse, Albuquerque

N. Dean Ricer, Park Superintendent,
Living Desert, Carlsbad

Dr. Norman Luonds, NMSU
Botanical Garden, Las Cruces

Special thanks to:
Maxine Johnson, Palm Desert, CA
and
Michele V. Millard, Tucson, AZ

And to:
Ronald Baetz, Desert Water Agency,
Palm Springs, CA

Francis Ching, Director Emmeritis,
LA County Arboretum

Cliff Douglas, Arid Zone Trees,
Queen Creek, AZ

Mary Rose Duffield, Tucson, AZ

Ron Gass, Mountain States
Nursery, Phoenix, AZ

Ron Gregory, Landscape Architect,
Palm Desert, CA

Warren D. Jones, Landscape
Architect, Tucson, AZ

Fred Lang, Landscape Architect,
Laguna Beach, CA

Michael MacCaskey, *Sunset*
Magazine, Los Angeles, CA

Luann Munns, LA County
Arboretum, Arcadia, CA

Bette Nesbitt, horticulturist,
Tucson, AZ

Robert Perry, Landscape Architect,
LaVerne, CA

W.G. Scotty Scott, landscape
consultant, La Quinta, CA

Ken Smith, Landscape Architect,
Newberry Park, CA

Merrill Windsor, editor, Phoenix, AZ

Sally Wasowski, Dallas, TX

Ruth Watling, landscape consultant,
Palm Desert, CA

Table of Contents

GARDEN DELIGHTS

Visiting the beautiful gardens of the Southwest becomes a series of wonderful discoveries. Like so many treasures, gardens of all sorts and sizes dot the landscape of Southern California, Arizona, Nevada and New Mexico. Some can be found in large cities, set back on busy streets. Other gardens are part of a college campus, city park, private business or working nursery, requiring a bit of diligence to uncover. But, as you'll see in these pages, the rewards are well worth the effort.

Today, more than ever, these gardens provide their community, their visitors and the world with a wealth of benefits. They offer education and awareness on a range of subjects, to everyone from the resident horticultural expert to the out-of-state visitor. Their landscaped grounds offer a respite from a fast-paced world: Pass through garden gates and you immediately enter an environment to enjoy and embrace. Gardens also play an important part in everyone's future—becoming refuges for plants whose habitats are under seige. In a sense, they are the zoos of the plant world, providing a place to nurture plants in danger of extinction.

A common thread of many gardens in this book is a vital issue to the West—*water conservation*. Botanical gardens and arboretums are proving grounds for plants and gardening practices to assist both the home and commercial landscaper in ways to reduce water use. Special emphasis has been given to these gardens and their conservation efforts.

The goal of this book is to provide much-deserved exposure to these valuable, regional resources. We hope it will help you make some discoveries of your own—whether in your neighborhood or during your next out-of-town excursion. Even if you don't get the opportunity to visit a single garden, we think you'll enjoy reading and learning about these gardens of the West—their history, the trail-by-trail accounts and vivid photographic descriptions.

Doing it for the beauty: A garden scene dominated by California poppies, Eschscholzia californica, at South Coast Botanical Garden in Palos Verdes Estates, illustrates just one reason to visit the gardens of the West.

How To Use This Book

The following pages describe 54 gardens located in Southern California, Arizona, Nevada and New Mexico. "Gardens" is used as a general term; most are much more, as you'll discover.

Each description is designed to provide practical, easy-to-access information, including directions to the garden, hours, whether fees are required and other basics. Every effort has been made to supply accurate, up-to-date information, but operating procedures for facilities do change. For this reason, we have left out exact admission prices, which tend to change most often. The conditions of a garden can also fluctuate rapidly due to the influence of the weather or season, or due to remodeling or expansion. It's best to call ahead if you are not certain of a garden's appearance, especially if you are traveling any distance.

In-depth profiles of the gardens—their history, descriptions of the special gardens and plants on the grounds—are discussed in the narrative sections. Many serve as a "walking tour." This information, in conjunction with the color photographs and illustrations, could be the next best thing to being there.

Botanical Garden, Arboretum. . .What's in a Name?

You'll notice that some gardens are listed as arboretums, others as botanical gardens or demonstration gardens. These basic definitions provide an overview of the different kinds of gardens discussed in this book.

Arboretum—Taken literally, an *arboretum* is a collection of living trees. Currently, the term denotes a plantation of many kinds of woody plants permanently maintained for purposes of study, rescarch and education, as distinct from a grove, forest, nursery or park. An arboretum is often a part of a botanical garden.

The Desert Botanical Garden in Phoenix is a prime example of a botanical garden, serving as a learning center and attraction for residents as well as seasonal visitors.

Botanical Gardens—A botanical garden is a controlled, staffed institution that maintains a living collection of plants. Education and research, in combination with libraries, herbaria laboratories and museums are goals. Each botanical garden develops its own special field of interest, depending on personnel, location, extent, available funds and terms of its charter. Greenhouses, test grounds, demonstration gardens, an herbarium, an arboretum and other departments are often included. The most important element is acquiring knowledge and making it available.

Demonstration Gardens—These practical gardens showcase selected aspects of plants and gardening. Emphasis is on take-home ideas. In the West, water-conservation gardens are most common, with water purveyors the leaders in establishing gardens. Botanic gardens and arboretums often develop their own demo gardens.

Private Estates—These are usually home gardens formerly owned by an individual then donated to a city or state. They are often managed by endowments and foundations. Private estates are reflections of horticultural history and allow visitors to appreciate the efforts of past generations of landscape designers, plant collectors and nurserymen.

Rancho Los Alamitos in Long Beach was once a 28,500-acre ranch. Today, its 7-1/2 acres of gardens are enjoyed by thousands of visitors each year. Trees planted by ranch owners during the 1870s still remain.

Zoological Gardens—These are gardens that also feature animal life. One of the best examples is the San Diego Zoo. These facilities are developing into thoughtfully landscaped *bioclimatic zones*—natural environments for their inhabitants. (For more on this, see page 50.)

National and State Parks—These natural landscapes offer a valuable sense of space. An entire book could be written on this subject, so our listings are by no means conclusive. We've included a few favorites.

Commercial Gardens—Many commercial plant growers and retailers develop gardens—to test plant performance or to showcase their wares. Some are as beautiful and appealing as botanical gardens. You'll find listings of these gardens at the conclusion of state-by-state sections.

Garden Climates

One of the fascinating aspects of visiting the gardens in Southern California and the Southwest is the amazing variety of plants and climates in a relatively small region. The influence of the Pacific Ocean, inland deserts, mountain ranges, rainfall, humidity, wind and variable soils combine to create a wide range of growing conditions, determining which plants will thrive.

In fact, botanical gardens are living, working examples of the many climates of the West. You can see how the plant palette changes from garden to garden, and how climate affects a garden's appearance. Compare, for example, gardens in Southern California. Tropical plants thrive in coastal gardens, yet are absent less than 40 miles inland, where more cold-hardy, temperate plants are grown. The contrast is more dramatic when desert regions are considered, where high temperatures and extremely low rainfall greatly change the spectrum of plants.

If you have the opportunity to visit several of the gardens in this book, be aware of the subtle and sometimes not so subtle effects climate has on each garden's range of adapted plants. Knowing a little about a garden's climate will help you understand why its appearance can be remarkably different from another garden just miles away.

Southern California Gardens

In this book, the gardens of Southern California are organized into three basic climate zones to reflect major climatic influences: *Coastal, Inland* and *Desert*. These are the primary factors that make these climates and gardens unique.

Many gardens located along the Southern California coast are blessed with mild temperatures and high humidity–optimum growing conditions for tropical and subtropical plants. This is Fern Canyon at the San Diego Zoo.

Southern California Coastal—The area along the Pacific Coast, generally extending five miles inland, is affected by ocean winds and fog. Summer sunshine and temperatures are moderate, and humidity is high compared to inland regions, which means plants need less water. The *growing season,* days between the first and last frost, is 350 to 365 days. Low temperatures are normally quite mild, and tropical and subtropical plants are often grown outdoors year-around. Rainfall averages range from 9½ inches a year in San Diego to 17 inches at Santa Barbara.

This is the climate of such lush, spectacular gardens as Balboa Park, the San Diego Zoo and Quail Gardens in nearby Encinitas. Others include the Hortense Miller Garden in Laguna Beach, and Sherman Library and Botanical Garden in Corona del Mar. A few miles inland, coastal influences remain strong at the University of California Irvine Botanic Garden. At the base of Palos Verdes Estates, north of Long Beach, is the South Coast Botanical Garden. Nearby, at Malibu, is the J. Paul Getty Museum. A few miles inland at UCLA are Hannah Carter Japanese Garden and Mildred Mathias Botanical Garden. Traveling north up the coast takes you to Santa Barbara Botanic Garden and Alice Keck Park, the northern-most point of the Southern California coastal climate. See pages 33 to 56.

Southern California Inland—This large climate zone could be further subdivided into another half-dozen smaller climates. The nature of the climates in this area largely depends on which has the greatest influence— the cool ocean or the hot, inland deserts. The hilly terrain mixes things up even more. Low temperatures are lower than along the coast, which makes it more difficult to grow tropical plants. High temperatures are also higher by comparison—100F days in summer are common. Humidity is

Gardens in inland valleys have different growing conditions compared to coastal gardens, and the look of the gardens can vary considerably. Shown is Rancho Santa Ana Botanic Garden in Claremont.

lower, but rainfall is greater—averaging 10 inches in Riverside to 18 inches in the San Gabriel Valley.

The look of gardens in this zone is more variable than coastal gardens. Fullerton Arboretum, about 18 miles from the coast, tends to be subtropical in nature, but is also influenced by warmer inland temperatures. Traveling only 20 miles farther inland to Rancho Santa Ana in Claremont, lower low temperatures and higher high temperatures means that more temperate, woody plants are grown. The same is true at LA County Arboretum in Arcadia, The Huntington Gardens in nearly San Marino and Descanso Gardens in La Canada. Farther inland, the University of California at Riverside Arboretum is beyond influence of ocean breezes. The combination of high heat and protection from low temperature extremes makes the UCR Arboretum prime citrus country. Southern California Inland gardens are described on pages 57 to 75.

Southern California Desert—Like inland regions, this area has a number of distinct climates, ranging from the low-elevation Sonoran Desert in Palm Springs, to the much colder high-elevation Mojave Desert, such as around Lancaster. In the low desert, the growing season is 325-350 days, and average annual rainfall is only 4 to 5 inches a year. In the Coachella, Imperial and Borrego Valleys, temperatures reach 110F-115F in summer and 26F-32F in late winter. Strong winds dessicate plants, requiring windbreaks or walls for protection. Subtropical plants will thrive if protected. Native, water-efficient plants are important in this environment.

Gardens in this region include the Living Desert in Palm Desert, Moorten Botanical Gardens and Desert Water Agency in Palm Springs. See pages 76 to 80.

Arizona

The Arizona gardens discussed in this book are a good representation of this state's climates, ranging from the cold, high-elevation climate at the Arboretum at Flagstaff to the Sonoran Desert in southern Arizona.

The Arboretum at Flagstaff is at an elevation of 7,150 feet and has a growing season (days between first and last frosts) of about 118 days. Cold temperatures decide which plants can be grown, and include mostly coniferous species of the Colorado Plateau. Average annual rainfall is about 20 inches.

The Sonoran Desert in the Phoenix area is at a 1,200-foot elevation, with an average of only 7 inches of rainfall each year. Temperatures *average* 105F in summer. This compares to Palm Springs' 110F-115F. The Desert Botanic Garden in Phoenix features desert natives, cacti, succulents and other dry-climate plants.

In the middle-elevation Sonoran Desert (2,200 feet) at Tucson, rainfall averages about 11½ inches a year. Practically half falls during the summer months. During a rainy summer, the gardens at Arizona Sonora Desert Museum, Tucson Botanical Gardens and Tohono Chul Park come alive with fresh, new growth and color, as plants quickly respond to the moisture. Summer temperatures are generally 5F below that of Phoenix. Winter low temperatures restrict the use of many cold-tender, subtropical plants to sheltered microclimates. Due to the higher elevation and increased rainfall, gardens are more lush than in other desert climates.

Boyce Thompson Southwest Arboretum in Superior, 125 miles north of Tucson, possesses a similar climate, but Magma Ridge and surrounding mountains temper conditions somewhat. For a discussion of Arizona gardens, see pages 14 to 31.

Nevada

Las Vegas and Henderson are located at the eastern section of the Mojave Desert. Their 2,000-foot elevation is similar to Tucson, but the growing conditions are more harsh, with greater temperature extremes. Annual rainfall is also much lower, averaging only 3 or 4 inches each year. Tough, cold winters, with temperatures in the mid-20s, requires use of cold-tolerant plants.

Gardens in this region include Ethel M Botanic Gardens in Henderson, the University of Nevada at Las Vegas' campus-wide arboretum and the Las Vegas Desert Demonstration Garden. See pages 88 to 91.

New Mexico

New Mexico is high desert country, with elevations ranging 3,300 to 4,500 feet and more. The climate of New Mexican gardens described in this book is similar to the high desert of California, but rainfall is slightly higher, from an average of 8 to 12 inches annually. Summer rains account for almost half. Winter temperatures drop down in the 20s; plants must be cold-hardy to survive.

The gardens in New Mexico are described on pages 92 to 95. They include the University of New Mexico Department of Biology greenhouse and the Albuquerque Garden Center in Albuquerque, the Living Desert Zoological and Botanical State Park in Carlsbad, and the New Mexico State University Botanical Garden in Las Cruces.

Southwest desert gardens have their own distinct appearance. These Teddy bear cholla cacti, Opuntia bigelovii, are located in the Ethel M Garden in Henderson, Nevada.

Map and Index of Gardens

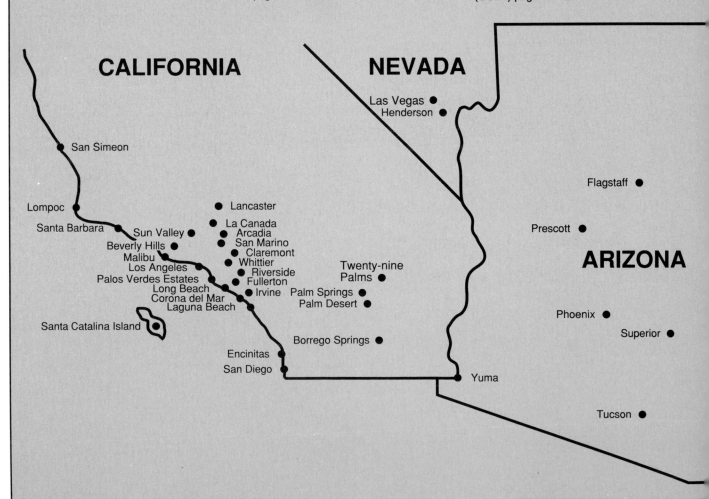

Albuquerque ●

NEW MEXICO

Las Cruces ●   Carlsbad ●

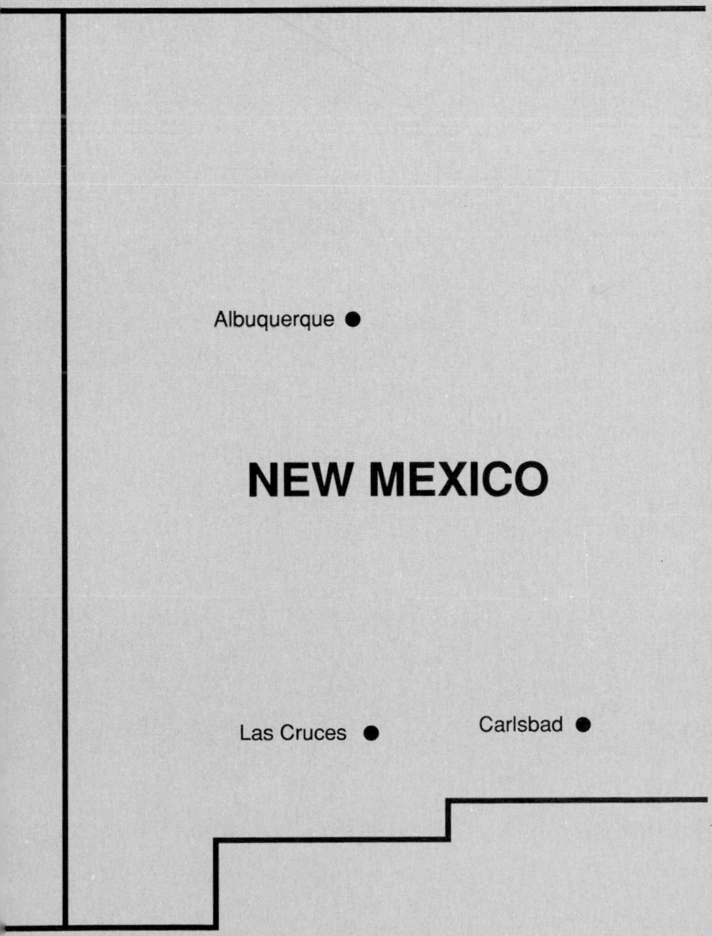

When You Visit . . .

☐ Wear comfortable shoes. Touring a garden usually involves lots of walking. Sometimes surfaces are uneven, with changes in elevation.

☐ Dress for the weather and the season. This could mean a jacket if you visit a coastal garden such as Hearst's Castle in San Simeon, or cool walking shorts at The Living Desert in Palm Desert. The weather can also change rapidly, so come prepared.

☐ Bring sun screen and a sun hat in warm areas. Even during winter, the sun in desert and inland regions can be intense. This protection will help you avoid sunburn and stress from the heat.

☐ Carry water with you. A container of water will be appreciated, even necessary, if water fountains are few and far between. Use your own judgement. Many gardens have plenty of water, but some don't, especially those with extensive nature trails.

☐ Stay alert while on trails. Most gardens are also home to a wide range of wildlife, some that can cause you harm if provoked. Leave animals in the wild alone. Also, be aware of sharp-pointed plants along paths, particularly in desert gardens. A close encounter with a cholla cactus could ruin your day.

☐ Leave your pets at home. Almost every garden prohibits pets, even dogs on leashes, on their grounds.

☐ Bring a camera! This is not a precaution, but a tip to make your visit more memorable. Capturing a photograph of a wildflower meadow or rose garden in peak bloom can be a real thrill.

ARIZONA
GARDENS

April at the Desert Botanic Garden in Phoenix.

The Arboretum at Flagstaff

Education is one of the primary goals of the Arboretum at Flagstaff. Visitors to the arboretum will gain an understanding of the plants and plant communities of the Colorado Plateau—located in northern Arizona, northeastern New Mexico, southern Colorado and southern Utah. Because this is a region of high elevation and generally low rainfall, the arboretum is involved in researching new water-efficient plants, those that have potential to be used as landscape ornamentals, as well as those that may have medicinal, food and fiber value. Additional research includes wind, solar, and water-harvesting techniques, as they relate to the production of plants.

For more down-to-earth subjects, home gardeners can gain knowledge from the many seminars offered by arboretum staff. A sampling of classes offered include: how to force bulbs, basics on soils, grafting, home greenhouses, how to extend the vegetable growing season and low-water landscape design and plant selection. Other broader-spectrum subjects have included classes on understanding the greenhouse effect, a primer on the geography of the Colorado Plateau and high-altitude horticulture.

Protecting Endangered Plants

One of the most important aspects of the arboretum is involvement in protecting rare and endangered plants. It is one of two gardens in Arizona (the other being Desert Botanical Garden in Phoenix) that is a member of the Center for Plant Conservation. This organization, created in 1984, is composed of botanical gardens and arboretums throughout the United States. Currently, there are 19 gardens nationwide, all dedicated to the preservation and study of all endangered American plants. To date, the Arboretum at Flagstaff is working with conserving 20 species of plants, studying their habitats, growth cycles and how they can be propagated. Plans are to expand this program.

Facilities at the arboretum include a 5,500 square-foot horticultural center set amid wooded surroundings. The center, contructed of native stone, also includes a solar greenhouse as well as offices and plant propagation areas. More than 520 plant species can be seen on the arboretum grounds.

Future plans are extensive. Plans for the next decade include a new visitors facility and several new specialized gardens, which combined will cover an additional 40 acres.

The Arboretum at Flagstaff
Mailing address: PO Box 670
Flagstaff, Arizona 86002
(602) 774-1441

From I-40 south of Flagstaff, exit at Woody Mountain Road. Travel south 3.8 miles to the arboretum entrance.

Over 200 acres of ponderosa pine forest land; approximately 10 developed acres.

Open Monday-Friday, 10 to 3 p.m. Open Saturday during June, July and August.

Tours at 11 a.m. and 1 p.m.

No fee required.

Group tours available. Call for reservations in advance.

Gift shop.

No picnic facilities.

Wheelchair access is limited.

Special events during the year include *Xeriscape Seminar,* late February; *Northern Arizona Garden Week Celebration,* in early May; *Horticulture Fair & Plant Sale,* late June; *Christmas Herb Sale,* early December.

"At an elevation of 7,150 feet, the Arboretum at Flagstaff is the highest arboretum in the United States doing horticultural research. Through integrated programs of plant collections, research, and public education, it studies and disseminates information concerning horticulture appropriate for the dry, high-elevation communities of the West."

Beard-tongue

Desert Botanical Garden
120 North Galvin Parkway
Phoenix, Arizona 85008
(602) 941-1225
(602) 941-2867: Wildflower Hotline,
in service March 1 through April 30.
Assists with location and timing of
wildflower displays in Arizona.

———

Located 8 miles east of the center of
metropolitan Phoenix. From I-10 take
east Van Buren Street to north Galvin
Parkway. From Scottsdale Road to
McDowell Road, then west on Galvin
Parkway.

———

145 acres of landscaped grounds.
October through April is the most
comfortable time to visit. May through
September is usually quite hot. Touring
time is one to three hours.

———

Open daily 9 a.m. to sunset. Open at 8
a.m. during July and August. Closed
Christmas Day.

———

Entrance fee.

———

Gift shop hours: 9 to 5 p.m.

———

Restaurant hours: open 10 to 4 p.m.
Located at the rear of Webster
Building.

———

No picnic facilities.

———

Wheelchair access.

———

Special annual events include: *Music
in the Garden,* held fall, winter and
spring on the Ullman Terrace; *Desert
Fest,* a celebration of the desert in
bloom and annual cacti sale, April
weekend; *Noche de las Luminanas,*
December holiday event. Numerous
classes are available on natural crafts,
landscaping and uses of desert plants.
Call (602) 941-1225 for information.

———

The Desert Botanical Garden

The 145-acre Desert Botanical Garden is one of ten botanical gardens in the United States accredited by the American Association of Museums. It is a major attraction and learning center for local residents, as well as a must-see attraction for thousands of seasonal visitors from all over the world.

Educating visitors about the rich diversity of plants and the ecology of the Sonoran Desert is the primary purpose of the Desert Botanical Garden. Visitors who take the time to notice the garden's "sense of place," will be exposed to the nuances of living and gardening in the arid Southwest.

Desert Botanical Garden Trail Guide

A well-organized, illustrated trail guide, available at the garden entrance, describes the major plant features of the garden. Before you begin your walk, study the map that identifies the major exhibits and buildings. For example, the John H. Rhuart Demonstration Garden presents practical applications and ideas for home landscaping, such as a ramada-covered patio, flowering desert shrubs and canopy-creating shade trees as well as vegetable and herb gardens.

Outside the Rhuart Garden, giant, multi-armed cardons, boojum trees and other giant cacti will, by their shapes and stature, command your attention. Cardon, *Pachycereus pringlei,* is often mistaken for the saguaro, *Carnegiea gigantea.* They are native to northwestern Mexico and grow 50 to 65 feet tall with numerous branches. Nearby is the "upside-down tree," the boojum tree, *Idria columnaris,* a relative of the ocotillo. In this area are the Webster Center, the Richter Library and Earle Herbarium, which combine to serve as the garden's educational and research center.

From this point, trails lead south up the hill to the Arizona Native Plant Trail, which takes you to the garden's highest point. Here you can view the entire grounds and the development in the surrounding Salt River Valley. North is Camelback Mountain near Scottsdale; to the east lie the rugged Superstition Mountains.

After descending this trail, turn right for a visit to the "Plants and People of the Sonoran Desert." This section offers a study of the ancient civilizations that inhabited the area. Exhibits show how former residents raised crops, built shelters and prepared food.

Heading north from the Plants and People exhibits, this area is landscaped with plants native to Australia. From here the trail takes you to two lath houses—the Succulent House and Cactus House. Information panels describe the differences between cacti and succulents. Extensive collections of agaves, yuccas and aloes are clustered around the houses. Particularly interesting are the spectacular aloes. Hummingbirds and insects are attracted to the red, orange, yellow or white flowers that bloom from spring into summer.

Some Notes About The Garden

One of the most-impressive features of the Desert Botanical Garden is the maturity of the plant collections. Opened to the public in 1939, its 145 acres now include over 10,000 desert plants, with more than 2,500 species. One-third of the plants on display are from the Sonoran Desert. Others are native to Mexico, South America and more exotic deserts of the world, such as Australia's Karoo, Africa's Sahara and Asia's Gobi.

Plants of the Gardens—The exemplary plantings at Desert Botanical Garden serve as inspiration for creative use of colorful, hardy water-saving plants. Areas throughout the grounds demonstrate ways to save

Lavender-blooming Penstemon combines with a green-barked palo verde tree along walkway. In the background is the Webster building.

Saguaro blossoms

energy through effective use of landscape plants: Shading walls and roofs with wide spreading canopy-shaped trees, such as blue palo verde, *Cercidium floridium*, palo brea, *Cercidium praecox*, mesquite, *Prosopis* species, and sweet acacia, *Acacia smallii*. Numerous annual wildflowers and perennials add seasonal color. Texture and form is provided by flowering shrubs such as the Texas rangers and orange firecracker, *Justicia spicigera*.

A Natural Environment For All

Because the plantings are dense, wildlife is plentiful. Squirrels, jackrabbits, lizards and desert tortoises, especially in cool morning hours, are plentiful. Bird-watchers can anticipate sighting native and migratory birds, such as the curious and busy roadrunner, Gambel's quail, jewel-like hummingbirds, and desert hawks that spiral overhead.

This is also a garden for children. Unraveling the mysteries of the desert will help them realize why water resources are so valuable and how they can live in tune with the arid environment.

A young visitor tries his hand grinding seeds at the Plants and People of the Sonoran Desert exhibit.

The Boyce Thompson Southwestern Arboretum

One of the most interesting aspects of the Boyce Thompson Southwestern Arboretum is its spectacular setting—nestled against the reddish cliffs of rugged Magma Ridge. The steep cliffs provide a dramatic backdrop for nature trails, demonstration gardens and naturalistic plantings.

Approximately 2,277 different plant species can be seen in gardens and along the more than two miles of nature trails. In fact, a highlight of a visit to Boyce Thompson is walking the loop trail that winds around Magma Ridge, with gardens and native vegetation on both sides of the cliffs.

A Walking Tour

As you enter the arboretum from the Visitors Center, you'll notice plants in containers on display along both sides of the walkway. Specimens in this exhibit are changed on a regular basis. Some plants go on display when in peak bloom; others are showcased because they are unusual or are not in view along the park trails. Most are labeled with a short description, providing insight into the plant's special qualities.

Demonstration Gardens—These are a short distance from the Visitors Center. Head south, then west, following signs on the trail. These gardens feature drought-tolerant plants adapted to grow in southern Arizona. Just south of the demonstration gardens is the picnic area for those who want to relax, have lunch and enjoy the view.

Following the trail from the demonstration gardens-picnic area takes you to the Smith Building, used for exhibits and interpretation. Twin greenhouses are attached at each end of the building: one contains exotic succulents from around the world; the other exotic cacti.

Just east of the Smith Building is the beginning of the newly completed Chihuahuan Desert Trail. For one-quarter mile, the trail leads visitors through areas planted with vegetation native to the Chihuahuan Desert. Mature specimens as well as new plant introductions are included. Some of the representative plants that can be seen include giant tree yuccas, blackbrush acacia and several varieties of Texas rain sage. It is interesting to compare and contrast the plants of Chihuahuan Desert to the Sonoran Desert, in which Boyce Thompson is located.

Continuing east takes you to the Main Loop Trail. Near the trail's beginning is the cactus gardens, with dozens of different kinds of cacti tucked into canyon crevices and amid rock outcroppings. Trails loop and wind throughout this area, offering many surprises, including the dramatic golden barrel, *Echinocactus grusonii*. Another favorite plant in this area is the boojum tree, *Idria columnaris*. This bizzare plant— you'll know one when you see one—looks something like a leggy, bottom-heavy palm

Boyce Thompson Southwestern Arboretum
Mailing address: PO Box AB
Superior, Arizona 85273
(602) 689-2811 for recorded information.
(602) 689-2723

Located 60 miles east of Phoenix, on Highway 60, 3 miles west of Superior.

Over 35 acres and two miles of nature trails that represent plants and gardens adapted to live in the Sonoran Desert of southern Arizona. Managed cooperatively by Arizona State Parks, the University of Arizona, and the not-for-profit Boyce Thompson Southwestern Arboretum.

Open daily 8 to 5 p.m. Closed Christmas Day.

Entrance fee required; children under 5 free with an adult.

Group tours available. Call for reservations one month in advance.

Gift shop hours: as the arboretum.

Picnic facilities located about 100 yards southwest of the Visitors Center.

Wheelchair access to sections of the arboretum.

Facilities available for weddings, parties or other events by arrangement.

Special events during the year include *Fruit & Vegetable Gardening Workshop*, third Saturday in January; *The Arid Land Plant Show*, first full weekend of April; *The Fall Landscaping Festival*, the second weekend in November.

"The Boyce Thompson Arboretum contains a beautiful garden and natural areas in a spectacular scenic setting. Visitors enjoy the peaceful atmosphere, clean air, the chance to hear the sounds of nature and observe small wildlife. Educational opportunities abound along trails, in specialty gardens, at the Visitors Center and Smith Building. Collections focus on arid-adapted plants from all over the world, including many cacti and other succulents."

Boojum tree

Rock-lined paths mark the way through the Boyce Thompson Southwest Arboretum cactus gardens, enhanced by the rugged Magma Ridge in the background.

tree that has lost its topknot, with short, spiky branches protruding from the trunk. When in bloom, during July, a spray of yellow flowers does appear at the top.

Continuing east, the trail goes past Ayer Lake, up near Picketpost House, then down and around the far northeast section of Magma Ridge. Heading back west takes you through a lush native riparian area, fed by Queen Creek. On this side of the ridge, the mostly shady path takes you past plantings of exotic trees and shrubs, including Chinese pistache, Chilean palo verde and dwarf pomegranate. Also on this side of the ridge is the Wing Memorial Garden, at the old Clevenger House—a house where part of the roof and walls are created from the cliffs of Magma Ridge. The Wing Garden includes a fabulous old-fashioned herb garden and many flowering plants, including roses that climb high up the cliff walls. A massive cat's claw vine practically blankets the house—its yellow blossoms most profuse in late spring.

The path continues west, then turns back north, passing through individual groves of palms and towering eucalyptus, natives of Australia. Some of these trees were planted in the late 1920s; they now reach to 70 feet high. Once out of these groves, you've completed the loop trail. Follow the signs and head back west to the Visitors Center, gift shop and nursery.

The Arizona-Sonora Desert Museum

Superlatives abound when describing the Arizona-Sonora Desert Museum. It is, in fact, a living museum of animals and plants, but a collection of several names is required to fully explain its contents: perhaps *Museum-Zoo-Botanical Garden-Earth Science-Natural-History Educational Facility-Sonoran Desert-Reserve*. Frequently listed as one of the world's best zoos, the Desert Museum is visited by more than 500,000 people each year, making it, along with the Grand Canyon, one of Arizona's premier visitor attractions.

The Desert Museum's appeal stems from the creative, naturalistic methods used to present animal and plant collections. The focus of the educational exhibits is the Sonoran Desert—its geology, plants and animals. Therein lies one of the Museum's greatest attributes; it has something for everyone. Gardeners, botanists, birders, animal lovers, rock hounds, amateur geologists and spelunkers—all are able to explore their area of interest.

In the animal exhibits, artificial rock enclosures make it seem as if you are observing the animals in desert canyons. Featured animals include bighorn sheep, mountain lions, bears, wolves, coyotes, coati (a relative of the raccoon), small cats, including ocelots, jaguarundi and bobcats, plus a prarie dog town and many others.

The otter and beaver exhibit features a see-through glass panel set below the water's surface provides a fish-eye view of their underwater antics. A large, walk-in aviary is a miniature, natural landscape housing over 40 species of birds, which allows observation as if you were in the wild.

Other creative exhibits are placed strategically around the Museum grounds. The Congdon Earth Sciences Center, named in memory of Stephen House Congdon, a museum trustee for many years, is an authentic replica of underground limestone caves. Here are realistic wet and dry limestone vaults that includes stalagmites, stalactites, ledges and flows. Other sections of the center include a mining exhibit, where tailings seeded with minerals by museum staff can be "excavated" by visitors for souvenirs.

Humble Beginnings

The museum opened its doors on Labor Day, 1952. Founded by William H. Carr and Authur N. Pack, the small museum staff began refurbishing abandoned adobe buildings located 14 miles west of Tucson. Although the facilities were rudimentary, the site—with its views of six mountain ranges nestled in a valley lush with saguaros and desert shrubs—was spectacular. Over the past four decades, following Carr's philosophy of creating exhibits that were interesting and educational, the Desert Museum has become one of the most-beloved zoos and botanical gardens in the world.

Arizona-Sonora Desert Museum
2021 North Kinney Road
Tucson, Arizona 85743
(602) 883-2702 for recorded information
(602) 883-1380

Located 14 miles west of downtown Tucson. Head west from I-10 on Speedway through Gates Pass, then north on Kinney Road to the Museum entrance.

Over 15 developed acres on a total of 186 acres, including natural habitat zoo, walk-in aviary, demonstration gardens, earth science exhibits and more. Over 300 plant species and 200 live animal species.

Open daily 8:30 to 5 p.m. from mid-September to mid-March. Open 7:30 to 6 p.m. from mid-March to mid-September. No tickets sold one hour before closing.

Entrance fee required; children under 6 free with an adult.

Group tours (reduced rates for 20 or more) are available.

Gift shop hours: summer 9:30 to 4:30 p.m.; winter 10:30 to 4:30 p.m.

Restaurant hours: Opens one hour after museum opens; closes one hour before museum closes.

No picnic facilities, but facilities available nearby at Tucson Mountain Park.

Wheelchair access to all exhibits; wheelchairs and strollers available free.

Wedding facilities available with prior arrangement.

"More than a museum, more than a zoo or botanical garden, the Desert Museum seeks to teach appreciation and conservation of the natural resources of the Sonoran Desert by its enlightening interpretation and displays. It expresses the ecosystem of one area: geology, plants and animals of the Sonoran Desert."

A family of ducks finds shelter beneath a canopy of a mesquite tree at the otter and beaver pond exhibit.

The Demonstration Desert Garden

The Demonstration Desert Garden is located in the southeast section of the Desert Museum's grounds. Opened in 1963, it was one of the first public gardens to provide ideas and information on how to develop creative, liveable landscapes in arid climates by using water-efficient plants.

Visitors can see how to create a lush, green landscape without high water bills; how walls and hedges can be used to capture the winter sun or block drying winds; and how to attract desirable wildlife such as butterflies and hummingbirds around their homes. Some unusual uses of common plants are shown, such as a formal box hedge of jojoba, a plant grown commercially for the oil squeezed from its beans.

The Desert Demonstration Garden is also a showcase for new horticultural introductions. Displaying these plants helps stimulate wider use in the landscape. For example, the Desert Museum was instrumental in introducing and popularizing such plants as Baja fairy duster, *Calliandra* species, red queen's wreath, *Antigon leptopus*, several small agaves and 'Desert Museum' palo verde, a hybrid palo verde with superior traits. Plant introductions to come include Mojave sage, *Salvia mohavensis*, snapdragon vine, *Maurandya antirrhiniflora*, and long-leaf morning-glory, *Ipomoea longifolia*.

In addition to the Demonstration Desert Garden, many cacti and other plants are located along trails and in other exhibits. For the unusual, stop by the hummingbird enclosure. It is a residential-style garden landscaped with more than 40 different species of plants that attract wild hummers.

Inside you'll find several species of native hummingbirds that you can observe up-close.

The Cactus and Succulent Garden features a comprehensive collection of regional species. Many of the animal exhibits throughout the grounds are actually simulated habitats that include plants appropriate for the animal's native community.

George L. Mountainlion

A trip to the Desert Museum would not be complete without mention of George L. Mountainlion. George was a resident cougar for many years and a favorite of museum visitors. After he passed away, Bill Carr wrote his epitaph, now engraved on a stone tablet and placed on the museum grounds, repeated at right.

"I George L. Mountainlion, freely give all sights and sounds of nature to those who have the grace to enjoy God-made beauty.

"To humans who are tired, worried or discouraged, I bequeath the silence, majesty and peace of our great American Desert.

"To those who walk the trails, I bequeath the early morning voices of the birds and the glory of the flowering desert in the springtime.

"To the children who have enjoyed me, hearing me purr and watching me turn my somersaults, I offer the precious gift of laughter and joy. The world so needs these things.

"And lastly, I bequeath my own happy spirit and affection for others to all who may remember me and my museum, where I did my best to show people that I truly liked them."

Many of the Desert Museum's animal enclosures are created from natural-appearing artificial rock. The result is an improved environment for the animals, which is also enjoyed by visitors.

Tohono Chul Park

Tohono Chul Park
Exhibit Hall, Gift Gallery & Tea Room
7366 North Paseo del Norte
Tucson, Arizona 85704
(602) 575-8468 for recorded
information
(602) 742-6455

From I-10, take Ina Road exit east to
North Paseo del Norte. Go north a
short distance to the garden entrance.

Over 400 plant species on 36 acres of
demonstration gardens and nature
trails, including many patios, ramadas
and special gardens. Ample parking.

Park open daily all year 7 a.m. to
sunset.

Donation suggested.

Gift shop hours: 9:30 to 5 p.m.
Monday-Saturday; 11 to 5 p.m.
Sunday.

Restaurant hours: 8 to 5 p.m. daily.
(602) 797-1711.

Exhibit House hours: 9:30 to 5 p.m.
Monday-Saturday; 11 to 5 p.m.
Sunday.

Limited picnic facilities.

Wheelchair access to most areas.

Guided tours for adult and children's
groups available: call for reservations.

Selected facilities available for
weddings or other special occasions
by reservation.

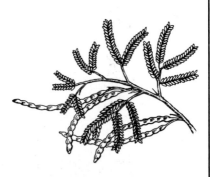

Native mesquite

One of the extraordinary things about Tohono Chul Park is its location—in the midst of a growing city. The park is a literal stone's throw away from one of Tucson's busiest intersections. Yet as you walk into the gardens, you soon forget the hum of nearby traffic. Well-groomed trails wind through an oasis of lush, natural desert growth to demonstration gardens, educational exhibits, gift shops and art gallery. Soon, your attention is captured by the tranquil beauty of this *tohono chul*—the Tohono O'odham Indian word that translates as *desert corner*.

Tohono Chul Park offers the visitor an island of natural beauty. It was established to promote the preservation of arid regions and to provide the public with an opportunity to learn about and experience the desert.

The park has two entrances. The main entrance is reached traveling north on Paseo del Norte from Ina Road. The other entrance is adjacent to The Haunted Bookshop, located just west of Oracle Road on north Northern Avenue. This bookstore, incidently, is also well worth a visit.

Paseo del Norte Entrance

Entering the grounds from the Paseo del Norte entrance, at the West House, you will find the Gift Gallery and Tea Room. The Tea Room is open for breakfast, lunch and afternoon tea. Drive or take the trails to the Entry Plaza complex, where the Exhibit Hall, another gift shop and administration offices are located. The Exhibit Hall features exhibits that focus on the desert and the Southwest, from bolla tie collections to Native American artifacts.

Northern Avenue Entrance

Near the Northern Avenue entrance, you'll find several demonstration gardens. Look, too, in this area for displays of yuccas, barrel cacti, succulents (the Succulent Ramada), hedgehog cacti, agave, cholla and prickly pear. A little farther down the trail going west, you'll also find the Pincushion Ramada, which houses about 150 species of *Mammillaria* cacti.

The demonstration gardens include several intimate shaded patios, providing visitors with ideas on how to use plants and materials around their own homes. A small, man-made stream simulates a riparian environment, recirculating water over and around huge, artificial rocks, nurturing water-loving plants. Additional plants, as well as native species of endangered fish such as the native desert pupfish, can be seen in the small rock grottos within the demonstration garden area.

Near this area you'll also discover the Geology Exhibit. This panoramic display illustrates the geologic history and development of the Santa Catalina Mountains, which can be seen by looking toward the northeast. A low-walled enclosure nearby houses several reptiles, including the desert tortoise, collared lizard and chuckwalla. Look for them in the early morning hours from April through October. Just north is the pottery wall, a unique, waterfall-fountain area. The greenhouse is nearby, visitors can purchase plants grown at the gardens. Greenhouse hours vary according to the season: Monday through Saturday.

East of this complex is the Ethnobotanic Garden. This display features various plants that have historical significance—plants grown by Southwest native people for food, fiber, medicine and dyes. Just north is one of the many birding areas, one of the park's most-popular attractions. Over 40 bird species find shelter and food on the park's 34 acres. Come early in the morning and you might catch a glimpse of a shy covey of quail, or a fleet-footed roadrunner.

Red-blooming fairy duster, Calliandra eriophylla, is graceful in combination with an ocotillo-rib window. Below: Demonstration gardens at Tohono Chul Park present creative ideas for landscapes, from plants to fountains to patio paving.

Tucson Botanical Gardens
2150 North Alvernon Way
Tucson, Arizona 85712
(602) 326-9255

Located in central Tucson, on North Alvernon Way just south of Grant Road. Take I-10 exit east on Grant, travel to Alvernon, then head south a short distance to entrance.

Over five acres of gardens and display areas, with over 400 plant species.

Open daily 8:30 to 4:30 p.m. Closed July 4th, Thanksgiving, Christmas Eve, Christmas Day, New Year's Day.

Entrance fee required; children under 12 free with adult.

Group tours available. Make reservations at least two weeks in advance.

Gift shop open 9 to 4 p.m. Monday-Friday; 10 to 4 p.m. Saturday; 12 to 4 p.m. Sunday.

Library hours: Tuesday 9 to 12 p.m. Contains over 3,000 volumes.

Picnic facilities for individuals (no groups).

Wheelchair access.

Facilities available for weddings, parties or other special events by reservation.

Special events during the year include *Spring Plant Sale*, March; *The Home Garden Tour*, April; *The Herb Fair*, July; *Fall Plant Sale* and *The Chile Fiesta*, both in October; and *Luminaria Night*, in December.

"Tucson Botanical Gardens provides a collection of gardens that demonstrates the horticultural possibilities in the Tucson area, as well as programs about plants for children, adults and the disabled. The gardens views itself as an educational institution, as well as a peaceful oasis in the midst of an urban area."

Tucson Botanical Gardens

The Tucson Botanical Gardens began as the beautifully landscaped home of the late Bernice and Rutger Porter, who operated a well-known plant nursery in Tucson for many years. Through the efforts of Harrison Yocum, Jr., and Lillian Fisher, the gardens were founded in 1968. In 1975, an additional 2½ acres was added, doubling its size. Approximately 400 plant species can be seen on the grounds, with more planned for the near future.

A Walking Tour of the Gardens

Although Tucson Botanical Gardens covers only 5 acres, small, special gardens and plantings are located throughout the garden grounds. Near the entrance, at the Porter House, you'll discover a kitchen herb garden tucked into a corner, planting squares defined with brick. Just south of the Porter House is the Reception Garden. This private spot, sheltered by an oleander hedge and flowering trees, is used often for weddings and other special events.

Walking south and east, past the fountain area, you'll come to a small herb-drying shed, where herbs grown on the grounds are cured. Following the winding walk, continuing east past the greenhouse filled with tropical plants, takes you to the Sensory Garden. This garden is actually a connected series of five separate gardens—each one specially designed to heighten each of the five senses. When completed, plantings will invite you to explore them via your touch, sight, smell, taste and hearing.

North of the Sensory Garden are the Heritage Rose Garden and Spring Wildflower Garden. Just north of these is the Growing Connections International Garden of the Child. The Heritage Rose Garden is a small garden featuring several of the original parent roses from which today's thousands of hybrids were developed. The Spring Wildflower Garden, best visited March through April, is planted with a mixture of Southwestern favorites, including gallardia, bright yellow desert marigold and California poppies. The International Garden of the Child features various ethnic plantings and promotes good nutrition through vegetable gardening.

Other gardens of note include the Native Seeds/S.E.A.R.C.H. demonstration gardens, which show time-honored techniques of growing native food crops. The Iris garden is just east of the entrance, facing the parking lot, and, nearby, the container garden, which includes several plant species planted in half-whiskey barrels.

Mexican evening primrose

The new xeriscape garden at Tucson Botanical Gardens.

Fresh chiles for sale at the garden's annual Chile Fiesta.

Xeriscape Garden and Solar Demonstration Garden

This is a recent addition, and one of the most exciting. (*Sunset* Magazine has followed the garden's progress in the pages of its magazine.) "Xeri-" translated from its Greek heritage, means "dry." When combined with "scape," Xeriscape becomes *dry landscape*.

This garden is a visual treat—with over 100 plant species—many which produce flowers thoughout the year. (The best time to visit, color-wise, is during late April to May.) The garden also gains high marks for its ability to educate. It teaches, by example, the ideas and techniques required to create an attractive, water-efficient landscape, such as grouping plants by water need into three zones: The Mini-Oasis, The Transition Zone and The Outer Zone.

A solar portion of the Xeriscape garden allows Tucson Botanical to take advantage of Tucson's 360 days of sunshine each year. Using solar collectors, the sun charges batteries, which power a fan and lights on the entry ramada, the timer for the irrigation system and the garden's lighting system.

The Xeriscape garden is sponsored by *SAWARA*, Southern Arizona Water Resources Association, and was created with the assistance of dozens of volunteers. Supporting literature and plant lists are available at the garden.

Additional Gardens of Arizona

MESA

Arizona Temple Garden
525 East Main
Mesa, Arizona 85203
(602) 964-7164

Mesa is located 16 miles southeast of Phoenix. Drive south from Phoenix on I-10, turn east on The Superstition Freeway (360). Take Mesa Drive exit, go north 3 miles to Main Street, travel east two blocks to the Temple.

Over 21 acres of gardens featuring over 26,000 bedding plants and 9 acres of lawn.

Open daily from 9 to 9 p.m. Tours of facilities available every half-hour.

No entrance fee.

Picnic facilities in nearby park.

Wheelchair access.

Special events include an *Easter Pageant,* the five nights preceding the holiday; *Christmas lights,* 275,000 of them, adorn the Visitors Center building, palm trees and shrubs during the month of December, when a million visitors enjoy the gardens.

The Arizona Temple Gardens of the Church of Jesus Christ of Latter Day Saints encompass 21 acres of impeccably groomed lawns, flowering annuals, stately palm trees and reflection ponds. One of the most appealing aspects of this garden is the cooling oasis effect it provides—in stark contrast to the surrounding Sonoran Desert. And if you still hanker for a dose of the desert, you can visit the Arizona cactus garden on the Temple grounds.

The LDS Temple is located at the center of the grounds. The nearby Visitors Center also has several pieces of religious art and sculpture on display. Guided tours are available every day of the year.

PHOENIX

Valley Garden Center, Phoenix
1809 North 15th Avenue
Phoenix, Arizona 85007
(602) 252-2120

From I-17 take McDowell Road east and turn north on Fifteenth Avenue. Located one-half mile north of West McDowell Road and West Palm Lane adjacent to Encanto Park.

The Valley Garden Center demonstrates many fine examples of home garden situations and plants adapted to the Salt River Valley environment.

Open 8 to 5 p.m. Tuesday-Saturday.

Guided group tours available by reservation.

No entrance fee.

Wheelchair access.

Facilities available for social events, weddings or meetings. Make reservations in advance.

Annual events include garden club sponsored shows. Call (602) 252-2120 for event dates.

The Valley Garden Center provides information and practical examples for Phoenix-area gardeners.

The Valley Garden Center is located in the middle of the City of Phoenix. It functions as an educational and social center for 16 garden club societies, as well as a site for public meetings of the greater Phoenix area. The clubhouse is a focal point, surrounded by eye-catching plantings of annuals and perennials. Begin your tour here before visiting the various demonstration gardens.

The center has a garden for almost every horticultural interest. For example, you'll find a well-planned, 1-acre rose garden of 1,125 plants (accredited in 1984 by the All-America Rose Society Public Garden Committee), a typical residential backyard garden with a cascading pool, a Japanese garden with koi fish, a tropical mini-oasis and garden of Southwest native plants, and a fruit tree orchard with citrus, loquats and stone fruit (such as peaches) adapted to the desert environment.

Garden club members, business firms and individuals donate their time and expertise to help home gardeners who live in the greater Phoenix area. For example, the center is a valuable source for information on water-conserving landscaping.

The trees planted throughout the garden serve as examples of attractive plants that thrive in a desert environment. They include flowering jacaranda (spring-blooming), crape myrtle (summer-blooming), orchid tree (late spring-blooming), and silk-floss tree (fall-blooming).

Native trees in the garden include mesquite, acacia, palo verde and featherbush. Dry climate shrubs such as the cassias, Texas ranger and bird-of-paradise blend with many other colorful and hardy plants.

PRESCOTT

Sharlot Hall Museum
415 West Gurley Street
Prescott, Arizona 86301
(602) 445-3122

Located in downtown Prescott, two blocks west of the square.

Grounds encompass one city block and include historical buildings and gardens that represent a slice of Southwest history, beginning with the founding of Prescott in 1864.

Tuesday to Saturday 10 to 5 p.m. April 1-Oct. 31; 10 to 4 p.m. November 1-March 31; Sunday 1 to 5 p.m.

Closed Mondays and major holidays.

Donation requested.

Gift shop hours: as the museum hours.

Picnic facilities available.

Wheelchair access.

Facilities available for weddings, parties or other special events by prior arrangement.

Special events during the year include *Museum Folk Art Fair*, the first weekend in June; *Fall Music Festival*, first weekend in October; and a lecture series in the fall: "Tea & Talk."

The namesake of Sharlot Hall Museum, Miss Sharlot Hall, was a collector and writer of Arizona history. Miss Hall gathered artifacts from the pioneers of Arizona and wrote about the colorful Indians, miners, ranchers and politicians of her time. In 1928, while territorial historian, her collection of artifacts was moved into the old governer's mansion and opened as a museum. Other major buildings with supporting artifacts include the John C. Fremont house, fifth territorial governor of Arizona, a Victorian-style home, the Bashford house and the Sharlot Hall building, which houses most of the exhibits. You can also step back in time to see a blacksmith shop, schoolhouse and ranch house.

Gardens at Sharlot Hall include a rose garden, where over 350 roses are planted in honor of famous Arizona women. Look, too, for a French Boursault rose, brought to Prescott in 1866 by the territorial governor's wife and planted at the governor's mansion. A pioneer herb garden and heirloom vegetable garden are also nearby. Gardens featuring native plants and endangered plant species will be added in the future.

One of nature's great swimming holes can be found in Tucson's Sabino Canyon.

The vegetable demonstration garden at Sharlot Hall Museum at Prescott.

TUCSON

Sabino Canyon Recreation Area
5900 N. Sabino Canyon Road
Tucson, Arizona 85715
(602) 749-8700
also
Sabino Canyon Tours, Inc.
(602) 749-2861 for recorded information.
(602) 749-2327 for Moonlight Tour and group tour reservations.

Located in northeast Tucson. From I-10, take Ina Road east to Skyline, which runs into Sunrise Road. Travel southeast to Sabino Canyon Road. Turn north (left) a short distance to entrance.

Park open daily, never closes.

No entrance fee.

Narrated shuttle bus tours available for a fee. Trams operate on the hour from 9 to 4 p.m. Monday-Friday. Saturday, Sunday and holidays they operate every half-hour 9 to 4:30 p.m. Moonlight tours three nights a month (when the moon is full), April through December. Call to reserve about two weeks in advance.

Visitors Center open daily (except Christmas Day) 8 to 4:30 p.m. Books, maps and postcards are available for purchase.

Picnic facilities abound.

Wheelchair access.

No pets allowed in the canyon.

Sabino Canyon recreation area is operated under the jurisdiction of Coronado National Forest. In addition, a private company, Sabino Canyon Tours, Inc., operates a tram service in the canyon, traveling the 3.8 miles of twisting, scenic roads up into the Santa Catalina Mountains. The road crosses over Sabino Creek several times on bridges built by the Civilian Conservation Corps (the C.C.C.) in the 1930s. The tour is an informative, Disneyland-kind of ride into a gorgeous riparian woodland surrounded by rugged, saguaro-studded cliffs. Riders can choose to simply take the 45-minute ride up and back, or get off at one of the nine stops—to hike or have a picnic—then catch one of the tram buses back down the hill.

Soldiers enjoyed Sabino's swimming holes back in the 1870s, and they remain popular today. Additional attractions are the wildlife, seasonal fall color, and cool relief from the surrounding Sonoran Desert. Cottonwoods, sycamores and native Arizona ash provide abundant shade along Sabino Creek.

YUMA

Yuma Century House Museum and Garden
240 Madison Avenue
Yuma, Arizona
(602) 782-1841

From I-8, exit at 4th Avenue, turn left on 1st Street, continue 3 blocks to Madison Ave.

This period house reflects the architectural features and gardens of a turn of the century home in the Southwest desert.

Open Tuesday-Saturday, 10 to 4 p.m. Closed Sunday, Monday and state holidays.

Group tours require reservations, individual visitations need no reservations.

Wheelchair access to home; paved garden areas are rough.

Gift shop hours: as museum hours.

Gardens are available for weddings, parties and meetings. Call for information.

The Century House and garden were built in the early 1870s and changed hands several times before developer E.F. Sanquinetti purchased the property in 1890. His goal of developing Yuma County along the banks of the Colorado River included the use of the Century House and a garden estate to entertain local citizens and businessmen.

Sanquinetti was determined to grow subtropicals such as hibiscus, bananas, palms and bamboo—as well as desert cacti and roses—to show clients the potential beauty in the unforgiving, near-rainless desert of Yuma, Arizona. These plants became important enhancements for Sanquinetti's aviaries, which housed exotic and local birds.

Donated in 1963 to the Yuma County Historical Society, the house and gardens became part of the Arizona Historical Society in 1981. Today, Century House and garden are headquarters of the Rio Colorado Chapter of the Arizona Historical Society. Eventually, the Century House, now a history museum, will be developed as a period house. The gardens will be renovated to reflect the period that will be interpreted.

Commercial Gardens

PHOENIX

Shepard Iris Garden
3342 W. Orangewood
Phoenix, AZ 85051
(602) 841-1231

Large selection of irises—spectacular when in bloom. Garden is open Thursday and Sundays during April.

TUCSON

B&B Cactus Farm
11550 E. Speedway Blvd.
Tucson, AZ 85748
(602) 721-4687

More than 600 species of cacti and succulents on display, covering 3.3 acres. Walk-through desert gardens, four greenhouses and showroom open 9 to 5 p.m. Monday-Saturday.

SOUTHERN CALIFORNIA
GARDENS

The English garden at South Coast Botanic Garden.

Sherman Library and Gardens

Covering a city block, Sherman Library and Gardens is a well-designed and immaculately maintained collection of colorful gardens and conservatories. The botanical offerings are varied and include more than 1,000 species and over 200 genera. Plants originate from around the globe, from desert regions to exotic tropical climates. The atmosphere at Sherman is intimate and relaxing, reinforced by the soothing sounds of fountains, waterfalls and tree branches moving in the coastal breezes. Terracotta brick walks surrounded by expanses of lawn and flowers help complete this idyllic setting.

One of the most-appealing aspects of Sherman is the luxurious use of garden color. As soon as you walk through the gates—literally yards away from the busy Pacific Coast Highway—your attention is captured by the expansive use of flowering plants. Fuchsias, flowering annuals and begonias are specialties. Planters, large and small, always brim with seasonal annuals. In the spring and summer look for marigolds and petunias. In fall and winter poinsettias, poppies and cyclamens replace them to take center stage. Hanging baskets of geraniums, fuchsias and begonias swing in the breeze in the Display Shade Garden, located adjacent to the Tea Garden, to the right as you enter the grounds. The Specimen Shade Garden, also nearby, features impatiens, ferns and several species of begonias.

One of most spectacular features is the Tropical Conservatory. Waterfalls spill into small pools ringed with lava rock; brightly colored koi fish add their own hues. The lush, vivid greens of ferns, palms and philodendron compete for attention amid orchids, gesneriads, bromeliads and crotons. Plants thrive in the climatically controlled environment, where a strict range of temperatures and high humidity are maintained.

Other gardens on site include the Rose Garden, Cactus and Succulent Garden, and the Discovery Garden. This is the most recent addition to Sherman, designed especially for the blind. However, anyone will appreciate the plants that have been selected for their ability to heighten our senses of touch and smell. The island-like shape of the planting beds make it easy to navigate through the Discovery Garden, as well.

The Library

The library is housed in a restored adobe home built in the mid-1940s; the wide-spreading California pepper tree outside the entrance was planted when the house was built. The fact that "Library" is listed first in its name, Sherman Library and Gardens, provides a good indication of its extent and importance. The materials in the library focus on the study of the Pacific Southwest, particularly the dramatic changes that have occurred in the last 100 years. Approximately 15,000 books, extensive

Sherman Library and Gardens
2647 East Pacific Coast Highway
Corona del Mar, California 92625
(714) 673-2261

Located between Newport Beach and Laguna Beach on the Pacific Coast. Take the Costa Mesa Freeway (55) to Corona del Mar Freeway, (73), to Pacific Coast Highway and entrance. Street-side parking.

Horticultural display gardens, featuring colorful annuals, hanging basket displays, and tropical and subtropical plants. 15,000-volume library focuses on recent history of the Pacific Southwest.

Open daily 10:30 to 4 p.m. Closed Thanksgiving, Christmas and New Years Day.

Entrance fee required.

Tours are available; make reservations one month in advance.

Library open Monday-Friday 9 to 5 p.m.

Gift shop hours: As the garden hours.

Restaurant hours: 11 to 3 p.m., Saturday, Sunday and Monday, plus Tuesday during summer.

No picnic facilities.

Wheelchair access.

Facilities available for weddings, parties and meetings by reservation. Call for information.

Fuchsia

collections of maps and photographs, about 200,000 documents and more than 2,000 reels of microfilm (containing back files on several newspapers, including *The Los Angeles Times* since 1881) are on file. The library is open to anyone who has the need to use the resources.

Right: The Sherman Library is housed in a restored adobe building. The wide-spreading tree is a California pepper, Schinus molle, planted in the 1940s.
Below: The Tropical Conservatory is filled with an array of tropical flowering and foliage plants, waterfalls and colorful koi fish.

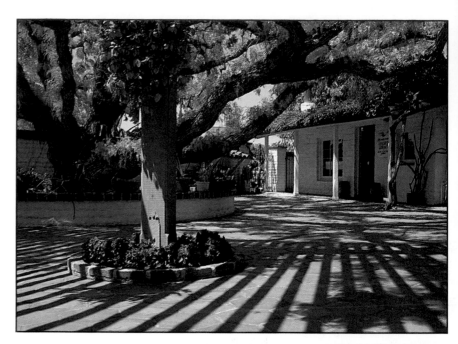

Quail Botanical Gardens

Located on the coast of California, just north of San Diego, Quail Botanical Gardens is blessed with a climate that allows a diverse assortment of plants to thrive. Frost-free growing conditions provide a home-away-from-home for tropical plants such as ferns, palms and uncommon fruit. Here you'll find, for example, the West's largest collection of hibiscus, as well as the largest grouping of bamboo species and varieties in the United States. Also sprinkled throughout the garden are numerous specimens of *Quercus suber*, commonly known as cork oak. You'll recognize its rough bark, which is used to make corks for bottles and cork products.

Quail's climate and location is also an ideal preserve for native California chapparal, once common to undeveloped hillsides in Southern California. An extensive Native Trail is located on the south side of the parking lot. A pamphlet is available describing how the area's earliest settlers used plants for survival.

Quail Botanical Gardens
230 Quail Gardens Drive
Encinitas, California 92024
(619) 436-3036
(619) 436-8301 Herbarium for plant information

Located about 20 miles north of San Diego. From I-5, take the Encinitas Blvd. exit, go east ½ mile to Quail Gardens Drive. Travel north ¼ mile to entrance.

Over 30 acres of landscapes and trails, gardens, pools and waterfalls.

Open daily 8 to 5 p.m.

No entrance fee required.

Nominal parking fee.

Group tours available by reservation at least one month in advance. Donation requested. Free tour every Saturday at 10 a.m. Meet at the gift shop.

Gift shop hours: Wednesday-Sunday 11 to 3 p.m.

No restaurant facilities.

Picnic facilities available.

Wheelchair access.

Facilities available for weddings (from April-October), parties or other special events by prior arrangement.

Highlights of annual events include: *Quail Botanical Gardens Foundation Biannual Plant Sales,* first weekend in May and first weekend in December; *Docent Society A Day in the Gardens Art & Photography Show & Plant Sale,* during October; *Quail Botanical Gardens Docent Society Summer Concert Under the Oaks,* late August.

The unusual, giant white angel's trumpet, Brugmansia versicolor, is just one of the fascinating plants you'll find at Quail Botanical Gardens.

Cork oak

Tropical plants thrive in the high humidity at base of the Mildred Macpherson Waterfall.

Gardens and Plant Collections

Quail Gardens has several theme gardens as well as numerous plant collections. Just north of the Visitors Center are the Desert Gardens. Continuing north will take you to the intimate Walled Garden, where weddings take place in the spring and summer months. North of the Walled Garden is a European herb garden and The Old-Fashioned Garden. Benches here are thoughtfully placed to face the Pacific Ocean, about a mile away. A demonstration garden featuring California natives has also recently been created. Its goal is to educate homeowners on ways to use natives in water-conserving landscapes.

One group of plants not usually included in public gardens is the Subtropical Fruit Garden, just west of this area. Many uncommon fruits are on display—some you may have never seen in cultivation before. Here are white sapote, feijoa, macadamia and cherimoya, as well the more familar avocado, kiwi, loquat and banana.

The frost-free climate allows a broad selection of plants to thrive in the Mildred Macpherson Waterfall and Palm Canyon area. Stairs and trails run parallel to the waterfall and along a recirculating stream. Lush plantings of palms, ferns and tropical plants create a jungle effect. The stream culminates in a boulder-lined pool brimming with waterlilies and flowering plants.

Plant collections—In addition to the extensive plantings of hibiscus and bamboo, sections are devoted to plants native to selected geographic regions. Most of these are located in the eastern and northern sections of the gardens. These collections include plants native to Central America, Australia, Himalaya, the South Pacific, South Africa and North America.

UC Irvine Arboretum

The University of California Irvine Arboretum began as a storage area for campus landscaping projects in 1964. Experimental gardens soon evolved, and the arboretum gradually expanded to include 10 acres. Today, the garden has matured and developed into a beautiful, enjoyable place where plants and people can mingle. Expansive areas of lawn combine with flowing walks to provide a pleasing sense of space.

At the entrance, pick up a copy of the guide brochure, which identifies the various gardens. As you enter, impressive collections of African succulents, cacti and American succulents immediately capture your attention. The diversity in size, color, shape and texture are extraordinary. The aloes are particularly striking, ranging in size from 6-inch miniatures to tree forms. Their brilliant, orange, yellow, and cream flowers bloom from spring through fall.

The arboretum has earned a reputation for its involvement in plant conservation. Almost 200 endangered species are maintained and actively propagated—many can be seen on the garden grounds. The arboretum also boasts one of the finest collections of South African bulbous and cormous plants. Several extensive landscaped borders along the walkways are devoted to amaryllis (125 species), iris (over 400 species), plus dozens of species of *Gladiolus, Babiana, Ixia, Moraea, Cyrtanthus* and *Lachenalia*. Mixed shrubs collected from other dry-climate regions represent the great variety of plants adapted to the Mediterranean climate of Southern California. Similar plants can be seen in the Sand Dune Display and South African Shrub Garden.

University of California Irvine
Arboretum
(North Campus)
Irvine, California 92717
(714) 856-5833

Take the 405 Freeway to Jamboree Road, head south to Campus Drive, go east less than ¼ mile to the first street (unnamed), and turn right to garden entrance.

Aloe gardens, cacti and succulents, South African shrubs and trees, sand dune displays and a great collection of bulbous and cormous plants.

Open 8 to 3:30 p.m. Monday-Friday. Closed on Saturday, Sunday and federal holidays.

No entrance fee.

Wheelchair access.

No picnic facilities.

Coral aloe

Flowering plants are abundant at the UC Irvine Arboretum. Shown are lavender-flowering orchid tree, Bauhinea variegata; yellow and orange Aloe species and the lily-like Hemerocallis.

The Hortense Miller Garden
Recreation and Social Services
Department
505 Forest Avenue
City of Laguna Beach, California
92651
(714) 497-0716

The garden is located at a private
residence in a gated community.
Guests are met at Riddle Field on
Hillcrest Drive in north Laguna and
escorted to the garden. All tours are
docent-led and must be arranged in
advance through the city's recreation
department.

Tours available every Wednesday and
Saturday, except for national holidays.
Tours are limited to a maximum of 15
guests per day and are scheduled from
10 until noon. Call for tour availability
and information.

No entrance fee.

No children allowed.

No picnic facilities.

No wheelchair access.

Ice plant

The Hortense Miller Garden

This is a beautiful, naturalized landscape that is full of surprises. It contains over 100 plant species integrated into a picturesque, 2½-acre setting located high above Laguna Beach, with view corridors to the blue Pacific Ocean. (The cover photo was taken at the Hortense Miller Garden.)

When visiting the Hortense Miller Garden, come prepared for exercise. Wear comfortable walking shoes; they are necessary to negotiate the many, steep, stepped paths that traverse the canyon slopes. Guides who conduct the tours beginning at the Allview Terrace entry are well-qualified to reveal the garden's hidden corners, and will help you seek out new plant treasures.

Coastal Climate Influences

Laguna Beach and the Hortense Miller Garden are situated in a unique climate, influenced by their proximity to the Pacific Ocean and numerous canyons. Plants are blessed with almost-ideal growing conditions. In addition to the mild climate, the canyon topography and shade created by the numerous mature trees and shrubs provide a selection of *microclimates*, small climates, taken advantage of by many resident plants.

Garden Origins

Well-known landscape architect, Fred Lang of Laguna Beach, worked with Mrs. Hortense Miller to develop the garden. He recalls Mrs. Miller's dedication and skillful use of plants to develop the well-blended natural landscape. Many coastal-adapted trees such as the Torrey pine, *Pinus torreyana*, canopy-creating oaks, lemon eucalyptus, *Eucalyptus citriodora*, coral trees and Australian willow myrtle, *Agonis flexuosa*, provide the basic structure and canopy. Numerous colorful shrubs and ground covers scramble over the ground and up and down the steep canyon slopes.

The garden began in 1950, and the first order of business was soil control on the steep slopes. This was accomplished with brilliant carpets of ice plants, tenacious zoysiagrass and colorful, trailing lantanas. In dry, sunny areas, shrubby, blue-flowering, California lilacs, Australian yellow senna, *Cassia*, and white-flowering sunroses, *Westringia*, tumble over rocks. Red bauhinia, *Bauhinia punctata*, also known as *Bauhinia galpinii*, produces color approaching that of bougainvillea.

A Garden Full of Surprises

One of the most dramatic experiments are clusters of pride of Madeira, *Echium fastuosum*, along a steep slope. Their eye-catching, blue, cone-shaped flowers are ideally suited to the sunny, coastal climate and low-water conditions.

Graceful, arching, bougainvillea add their brilliant, deep red shades. African aloes, with red, yellow and cream flowers, range in size from 6-inch miniatures to tall, forked species. Some species bloom every month, but most are more prolific from February through September. Agave clusters add unique plant forms.

Perennials, bulbous plants, African iris, calla lilies, ferns, giant lily turf, *Ophiopogon jaburan,* and lavender-flowering *Mackaya bella,* create a lush, colorful effect in the shady portions of the garden. Coastal natives are common in the large, northwest section. They demonstrate their vigor and ability to maintain growth even under drought conditions.

Bird-watchers will find the garden a haven for the wide variety of coastal species that visit during the year. Due to tour requirements, plan on at least two to three hours to fully enjoy the many beautiful aspects of this special, personal garden.

Much of the Hortense Miller garden is planted on steep slopes. Ice plant produces a brilliant display and protects against soil erosion.

Purple-flowering pride of Madeira creates a dramatic display alongside steps of railroad ties.

Rancho Los Alamitos

Rancho Los Alamitos Historic Site
6400 Bixby Hill Road
Long Beach, California 90815
(213) 431-3541

From the 405 Freeway in Long Beach exit at Palo Verde and head south. Enter through the Bixby Hills gated community to entrance.

Open Wednesday-Sunday, 1 to 5 p.m.

Tours of ranch house and barns are taken with docent and require 1-¼ hours. Gardens are self-guided, or with docent upon request. Special and group tours (exceeding 10 people), require a reservation.

Rancho Los Alamitos is one of the true, traditional California gardens of the late 18th and early 19th centuries. It is a garden that evolved without pretense in response to the land, climate, resources and the lives of many of its resident families.

No entrance fee.

Limited wheelchair access.

No picnic facilities.

Annual events include the *Herb Program*, July; *California Ranch Day*, first Sunday in June; *Rural Christmas*, second week in December. Monthly programs featuring domestic and agricultural skills of early 20th century and horticultural interests.

Jacarana

This is a garden that allows you to appreciate the past. Rancho Los Alamitos follows a different path than gardens copied from grand-scale estates of Italy, France and England, which were often expressions of wealth, power and status. Rancho Los Alamitos possesses an original, personal, unpretentious atmosphere that makes it a welcome, comfortable place to visit. The preservation of the garden is on-going, with the City of Long Beach and Rancho Los Alamitos Foundation cooperating to maintain this valuable historic site under the direction of landscape architect Russell Beatty and garden historian David Streatfield.

Historical Background

The ranch began as a land grant of 167,000 acres to Manuel Nieto in 1784. In 1834, the ranch was divided among the Nieto heirs, resulting in the creation of the 28,500-acre Rancho Los Alamitos. Today, the 7.5-acre historic site retains many of the trees planted from 1870 to 1890 by former ranch owners, the John Bixby family. In 1906, Fred Bixby inherited the ranch from his parents, Susan and John. Rancho Los Alamitos was transformed into a working horse ranch and headquarters for many other Bixby ranches.

Assisted by landscape architect Florence Yoch, Florence Bixby developed the gardens over a period of 30 years (1906-1935.) The area around the ranch house was gradually transformed into a series of intimate garden spaces, to be enjoyed by family and friends. Even today, the gardens retain their original forms as the Friendly Garden, Native Garden, Desert Garden, Rose Garden, Herb Garden and a small, walled Secret Garden. In the early 1930s, the Oleander Walk and Olive Patio were created to screen out urban sprawl prompted by the discovery of oil on the ranch and nearby.

A Garden Tour

From the parking lot off Bixby Hill Road, you begin the tour at the Reception Office, and a walk through the original, formal garden, largely developed in the 1890s. (Walking shoes are necessary—some pathways and surfaces are uneven.) You'll notice the stump remains of a massive, gnarled California pepper, *Schinus molle*, planted around 1840. The "old" garden, redesigned in 1922, is filled with fascinating details. A few examples are the fountain, with a backdrop of vertical accents of delicate, airy bamboo, and clumps of calla lilies at the base of tall, Italian cypress. As you move through the extensive California native cacti and succulent garden, you become aware of the thought given to the design initiated by Susan Bixby, her landscape architects and horticultural friends.

The formal long vista garden walks that lead you to other garden areas hold many surprises. After you leave the cactus gardens, note the

The Geranium Walk at Rancho Los Alamitos. At one time, this area was set aside for the resident children's pet goats, ponies, dogs and rabbits.

jacaranda trees, with their distinctive, lavender-blue blossoms, that line the walk just east of the tennis court. The Friendly Garden is nearby, where friends brought gifts of plants. The Oleander Walk leads to the Rose Garden; the Geranium Walk leads you back to the original front entry. This is where massive, Morton Bay fig trees extend their heavy, root systems over the ground and canopy of shade over the lawn.

Moving through the gardens, terraces, amid vine-covered walls and trellises, and the great variety of trees and plants, its difficult to imagine this historic property was once a ranch that became a horticultural dream come true. Russell Beatty aptly describes Rancho Los Alamitos as "An Island in Time."

One of the two massive, Morton Bay fig trees, planted more than 100 years ago, mark the original entrance to the Rancho Los Alamitos ranch house.

UCLA Hannah Carter Japanese
Garden
10619 Bellagio Road
Los Angeles, California 90077
(213) 825-4574 (UCLA Visitors
Center)

Take the 405 Freeway to Sunset Blvd.
Turn north to Stone Canyon Road and
proceed to the stop sign. Bear left to
Bellagio Road, then to the entrance—
just beyond the first two homes at the
intersection of Stone Canyon Road and
Bellagio Road.

A 1-½ acre garden that shows the
elements of a traditional Japanese
home life, garden plants in a natural
woodland setting and special
collections of dwarfed trees. All
combine with recurring asymmetric
forms and details to reflect the quest
for a quiet and peaceful retreat.

Garden can be seen by reservation
only: Tuesdays 10 to 1 p.m.;
Wednesdays 12 to 3 p.m.

No wheelchair access.

Japanese maple

The Hannah Carter Japanese Garden

Located a short distance from the University of California at Los Angeles (UCLA) campus, the Hannah Carter Japanese Garden is an intimate, traditional Japanese garden. Its key elements—water, stones, structures and plants—are woven together to create a memorable garden experience.

Garden History

The garden was created for its owners, Mr. and Mrs. Gordon Guiberson, in 1961. Both were fond of Japanese gardens and had studied them extensively. To ensure the garden's authenticity, they brought from Japan Nagao Sakurai, a leading landscape architect, to design the 1.5-acre parcel. In 1965, the garden was donated to UCLA by Edwin W. Carter, then Chairman of the Regents of the University, who had purchased the property from the Guibersons. The gardens received their current namesake, in honor of Carter's wife, Hannah.

The main gate, teahouse, bridges and shrine were built in Japan and re-assembled on-site by Japanese artisans. Primary symbolic rocks, antique stone carvings and water basins were also imported. Some 400 tons of dark brown stone were transported to the site from Santa Paula Canyon in Ventura County.

Each element in a Japanese garden possesses a symbolic role. The Hannah Carter Garden is a *chisen-style* garden—a pond or lake occupies important portions. In dry, *karasanus-style* gardens, patterns are raked in sand or gravel to simulate the sea. Stone groupings at the edge of pools represent a rocky seashore.

The Garden's Plants

Almost 100 plant species and varieties grace this garden. Traditionally, many Japanese plants have symbolic meanings that are closely interwoven with daily life. Most important are pines and bamboo, which express *longevity*. Flowering plum embodies the qualities of *vigor* and *patience*. Mature California live oaks, *Quercus agrifolia*, on the grounds before the garden was installed, add to the tranquil mood, plus provide an extensive canopy of shade. Of special interest are the Yulan magnolia trees, *Magnolia heptapeta*, located by the deck and below the teahouse. Nearby, four varieties of Japanese maple add a delicate touch of fall color. Black bamboo, *Phyllostachys nigra*, combine with eight other bamboo species to create special effects due to their arching, graceful growth patterns. Heavenly bamboo, *Nandina domestica*, adds a delicate, bamboo-like effect near the pool. Rhododendron, azaleas and camellias contribute late winter and early spring color. Giant lily turf, *Liriope gigantea*, and mondo grass, *Ophiopogon japonicus*, are tucked among rocks as accents and ground covers. Behind the teahouse, a dramatic waterfall helps make the Hawaiian Garden a dramatic attraction. Tropical plants here thrive on the additional moisture and humidity provided by the falls.

Mildred E. Mathias Botanical Garden

The Mildred E. Mathias Botanical Garden is located on the southeastern edge of the sprawling University of California Los Angeles campus. The climate is influenced by its proximity to the coast; frost seldom occurs. This allows a wide selection of subtropical and tropical plants to thrive here. Over 3,000 plant species and 205 plant families grow on the grounds. In addition to being home to these thousands of plants, the garden conducts horticultural research.

A Garden Tour

Each section of the garden offers unique discoveries. South of the Lath House, lily beds were developed in 1983. They contain true lilies and their relatives from all continents. For best bloom, visit during the spring months.

As you walk through the Australian section you will discover a planting of living fossils—*cycads*. About two dozen species of cycads grow in grassy areas shared with trees and shrubs of the legume family and several species of eucalyptus. On the bank east of the cycads is a collection of plants from the mountains of tropical Central America and South America.

Continuing south takes you to the Malesian rhododendron beds. These have been sponsored by the American Rhododendron Society. You will also find many other genera of the heath family, *Ericaceae*. All of the plants in this area require acid soil and good drainage.

The south section of the garden includes one of the finest collections of eucalyptus in Southern California. In addition to these Australian natives, you will see collections of *Leptospermum, Callistemon, Hakea, Grevillia* and *Acacia.*

As you continue north, note the *Gymnosperms*, which include conifers (pines, cedars), maidenhair trees, *Ginkgo biloba*, fern pine, *Podocarpus* species, and *Araucaria* species, including the monkey puzzle tree, and bunya-bunya. Some outstanding specimens of the dawn redwood, *Metasequoia glyptostroboides*, may be found near the stream. Notice, too, in the garden's center, the 25-plus species of palms, representative of every tropical region.

On the eastern slope are plants indigenous to coastal sage scrub and chaparral communities. Many plants originate from northern Baja California and the Channel Islands of Santa Catalina, San Clemente and Santa Barbara, and are seldom found in mainland gardens.

A collection of cacti and succulents is located along Hilgard Avenue. Species have been collected from desert areas all over the world. South of the desert garden is a section devoted to shrubs native to Mediterranean climate zones.

The Mildred E. Mathias
Botanical Garden
University of California at Los Angeles
Los Angeles, California 90024-1606
(213) 825-3620 or 825-2714
or 825-1260

Located in the southeastern section of the UCLA Campus. From the 405 Freeway, exit at Sunset Blvd. and travel east to Hilgard Avenue. Entrances at the corner of Le Conte and Hilgard Avenue South, and between the Botany and Medical buildings at Tiverton Drive and Circle Drive South.

Parking on and around the campus is limited. With the exception of metered stalls, permits are required in all campus parking areas. Daily parking permits can be purchased at parking information stations located at major campus entrances.

Open Monday-Friday 8 to 5 p.m.; Saturday and Sunday 8 to 4 p.m. Closed university holidays.

No entrance fee.

Wheelchair access.

Picnic facilities.

Dr. Mildred E. Mathias was director of the Botanical Garden from 1956 to 1974. She has made distinguished contributions to horticulture and education at the University of California at Los Angeles since 1947. Her influence on Southern California horticulture can be felt today—in the academic world as well as in exploration and plant research.

Maidenhair tree

J. Paul Getty Museum
17985 Pacific Coast Highway
Malibu, California 90265-5799

Mailing address: PO Box 2122
Santa Monica, California 90406
(213) 458-2003 for reservations
and information

A re-creation of an ancient Roman country house and surrounding gardens— the Villa dei Pariri of the ancient city Herculaneum in Italy— buried by the eruption of Mount Vesuvius in A.D. 79. Museum houses collections of Greek and Roman antiquities, pre-20th Century European paintings, drawings, sculpture, illuminated manuscripts, decorative arts, and 19th and 20th century European and American photographs.

Open 10 to 5 p.m. Tuesday-Sunday. Closed Mondays, July 4, Thanksgiving, Christmas and New Years Day.

Talks offered daily in the galleries, gardens and auditorium on a variety of subjects.

No entrance fee.

Important: Parking space is limited. If arriving by car, a parking reservation is required.

Bookstore hours: 10 to 4:45 p.m.

Restaurant hours: refreshments available 9:30 a.m. to 4:40 p.m.; lunch available from 11 a.m. to 2:30 p.m.

No picnic facilities.

Wheelchair access.

Laurel

The J. Paul Getty Museum

This is a one-of-a-kind museum, art gallery and collection of elegant gardens located on the shores of the Pacific in Malibu. As if this romantic location wasn't enough, the buildings and grounds are faithful re-creations of a Roman villa buried by an eruption of Mount Vesuvius centuries ago. The mood and appearance of the gardens is authentic, due to the similarities of the Southern California coast and that of the Bay of Naples. In fact, many seeds and bulbs were imported from Italy and grown at a local nursery before being planted at the museum.

The Getty Museum Gardens

Gardens were important to the Romans, and practically every home had a garden of some sort. Their outdoor spaces—predecessors of today's patios--were cool, shady areas used for relaxing, eating meals and recreation. Plants believed to be grown include a wide variety of herbs, fruit trees, evergreens, olive and grapes. A walk through the grounds at the museum will take you through these many gardens:

Main Peristyle Garden—This is a large, retangular garden. Its primary feature is the long, narrow pool placed in the garden's center. The mood is formal yet peaceful, accentuated by reproductions of bronze statues and busts copied from originals in Herculaneum. Plants surrounding the pool are shaped symmetrically, and include trimmed boxwood hedges, laurel, ivy, roses, oleander, iris and woolly yarrow. Two large arbors are covered with grapevines. Other plants include pomegranate, fan palms and acanthus.

The Herb Garden—This long, narrow garden is located directly west of the Main Peristyle Garden. More than 50 herbs grace the numerous, geometrically shaped planting beds. Here you'll find just about every kitchen herb imaginable, plus fruit trees, vegetables, grapevines and a grove of olives. Native rock and neatly raked gravel walks blend with the predominately blue-gray and blue-green colors of the plants.

The West Garden—This is the location of the Tea Room restaurant. Prominent features are two fountains—replicas of ones found at Pompeii and Herculaneum. Tall columns, terraces and planting beds filled with colorful annuals complete this formal scene.

The East Garden—This is a more intimate, enclosed garden. It also features two fountains—one a large circular basin in the garden's center, the other an intricate, mosaic grotto built into the garden wall. Sycamore trees add cooling shade.

Inner Peristyle Garden—A long and very narrow pool ringed with low-growing ivy is the focal point of this small, intimate, square-shaped garden. Five large bronze statues of women surround the pool. Boxwood hedges and plants sculpted into globes are kept low to the ground to

Rich in History and Art

The Getty Museum is an intricate study of history, art, architecture and design, deserving suitable time to reflect and consider all of its offerings. It is a place of immense detail and nuance—every color, every column, every statue and every plant—has been selected and placed carefully to evoke the authentic flavor of historic Rome. The collection of art is extensive. You'll find separate sections of the museum devoted to Antiquities, Decorative Arts, Drawings, Majolica and Glass, Manuscripts, Painting and Sculpture and Photographs.

For more detailed information on this jewel on the Pacific, refer to the book, *The J. Paul Getty Museum: Guide to the Villa and Its Gardens,* available from the Getty Museum.

The Main Peristyle Garden at the Getty Museum. The long, narrow pool is the primary feature, surrounded by symmetrically shaped plants and low boxwood hedges.

South Coast Botanical Garden
26300 Crenshaw Blvd.
Palos Verdes Peninsula, California
90274
(213) 544-6815

Located south of Los Angeles. Take the 405 Freeway to Crenshaw Blvd., travel south about five miles to the garden entrance.

Over 87 acres of gardens, demonstration plots, natural landscapes and a four-acre lake.

Open daily 9 to 5 p.m. Closed Christmas Day.

Entrance fee required.

Gift shop hours: as the garden hours.

No restaurant facilities.

Picnic facilities.

Wheelchair access.

Facilities available for weddings, parties or other special events by prior arrangement.

Highlights of annual events include *South Coast Camellia Show*, January; *Fiesta de Flores Show & Plant Sale*, mid-May; *Norris Theatre for the Performing Arts*, November; *Holiday in the Gardens and Christmas Concert*, December.

The vegetable demonstration garden at South Coast Botanical Garden is one of the most extensive anywhere. Produce from the garden is donated to a local charity.

South Coast Botanical Garden

To best appreciate a visit to South Coast Botanical Garden, you should have an understanding of its unusual past. As you stroll around the grounds, enjoying the profusion of flowering plants, the well-tended demonstration plots and special gardens, you'll never imagine what the property was like a few decades ago, before they were developed into the beautiful gardens they are today.

For 25 years (from 1929 to 1956), the area that is now South Coast Botanical Gardens was mined for *diatomite*, the fossilized remains of algae. This material, also called *diatomaceous earth*, is used commercially, such as in filtering systems for swimming pools. After more than one million tons of diatomite were removed, the open pit mine left behind was utilized as a sanitary landfill. Eventually, over three million tons of trash filled the mine pit, creating a layer of trash roughly 100 feet deep. A three-foot layer of topsoil was added, and planting of the garden began in 1961. Today, life has been renewed. Over 200,000 plants thrive, as well as numerous animals, including over 200 species of birds.

A Guide to Garden Highlights

There is a casual, understated elegance to South Coast Botanical Garden. You can choose to spend time in more formally designed gardens such as the English Garden or Herb Garden, or walk along natural trails and roads through plant collections as diverse as redwoods, maidenhair trees, eucalyptus and fruit trees.

The first garden section near the entrance is actually composed of several, well-maintained, individual gardens. Here you'll find the covered Shade Garden, Herb Garden, Flower Garden, Vegetable Garden, Bromeliad Garden and Rose Garden. Each offers color and interest through the seasons. The Vegetable Garden is quite extensive. Maintained by volunteers, the vegetable produce is donated to a local Meals-on-Wheels organization.

Heading into the grounds, travelling east (to the right) on the paved road, takes you to the Succulents & Cacti Garden, which includes hundreds of specimens from around the world. Nearby, to the left of the road, are coral trees, tree of the City of Los Angeles. Just beyond this section, and farther east into the garden are several California natives, as well as flowering and fruit-producing trees. Continuing farther east (still on the paved road) will take you to the redwoods. Two species—*Sequoia gigantea* and *Sequoia sempervirens*—are displayed. These are just a few decades old, but mature adapted trees can reach up to 350 feet high.

The English Garden—just beyond the redwood trees is a special treat— an old-fashioned, English perennial garden. (See photo, page 32.) The climbing red rose Paul's Scarlet Climber drapes over a trellis to mark the entrance. A mass of brilliant, magenta bougainvillea serves as a striking backdrop to the traditional white garden bench.

The paved road curves gently back north, looping around the four-acre, man-made lake in the valley below. Near here is the weather station (records rainfall and other climate statistics), and an extensive collection of eucalyptus trees. Continuing north will take you to the maidenhair trees and the conifers and pines collection. Heading west on the road takes you to the palms collection. Just past the palms you will begin to head back south to the entrance. Along this stretch you'll find the large *Ficus* collection, thriving in the cool coastal climate.

Coral tree blossom

A white garden bench against a backdrop of Bougainvillea creates an idyllic setting. The English garden is at left.

Balboa Park

Balboa Park
City of San Diego:
Parks & Recreation Central Division
Balboa Park Management Center
San Diego, California 92101
(619) 236-5717

Balboa Park is central to the city of San Diego, located just east of Lindbergh International Airport. If within a 5-mile radius, tune your radio to 1610 AM for information.

Over 1,200 acres of landscaped grounds, special gardens, museums, theatres and more.

Park is open daily. The Botanical Building is open 10 to 4 p.m. Tuesday-Sunday. Museum hours vary; call 239-9628 for information. Closed Monday, Thanksgiving, Christmas and New Years Day.

No entrance fee to the park, but most museums require a fee. Some museums have free admittance certain Tuesdays of each month under the program, *Free Days*. Call for schedule.

Free walking tours offered each Saturday at 10 a.m. Meet in front of the Botanical Building. Topics include *Heart of the Park* (1st Saturday of month;) *The Palm Walk*, (2nd Saturday,) *The Tree Walk*, (3rd Saturday) and *The Desert Walk*, (4th Saturday).

Several restaurants and concession stands: hours vary.

Gift shops in some of the museums.

Picnic facilities abound.

Wheelchair access to most sections of the park.

Weddings by permit. Parties and meeting facilities available. Contact Balboa Park Management Center: 236-5717.

Special annual events include the *Summer Twilight in the Park Concert Series*; *Kidz Arts Festival*, mid-October; *Fiesta de la Quadrille*, first weekend in November; and *Christmas on the Prado*, early December.

Labeled "the cultural heart of San Diego," Balboa Park evolved into one of the premier parks in the United States. To call it a park is to miss much of its rich heritage and cultural diversity—it offers activities from the sublime to the elegant—hosting a reported 14 million visitors each year.

History of Balboa Park

San Diego's founding fathers had the foresight to set aside 1,400 acres for a spectacular city park, but most of the land lay undeveloped for years. Kate Sessions, a resident horticulturist, saw the land as an opportunity. Her efforts would forever change the face of San Diego, and Balboa Park in particular. Sessions planted trees on the park's grounds in exchange for the use of acreage to raise plants for her private nursery. She brought in plants from San Francisco and exotics from Hawaii, Australia, South America and Mexico. Her research and experiments paid off, introducing new plants that remain a part of today's Southern California landscape. These included the matilija poppy, *Romneya coulteri*, new varieties of bouganvillea and California lilac, *Ceanothus cyaneus*. Many of the landmark plants that can be seen at Balboa Park and around San Diego were planted by Sessions: the *Arecastrum romanzoffianum* palms along Sixth Avenue, lemon scented gum, *Eucalyptus citriodora*, cork oak, *Quercus suber*, and several species of *Acacia*.

The Panama-California Exposition—Following Session's literal ground-breaking efforts, the Panama-California Exposition Garden Fair launched an even more ambitious planting and building phase. The exposition was to commemorate the opening of the Panama Canal, scheduled for January 1, 1915. (In fact, the park is named after Vasco Nunez de Balboa, the Spanish explorer who first sighted the Pacific Ocean.) By the time the Panama-California Exposition opened, the Garden Fair, as it was called, featured some two million plants of more than 1,200 species. Horticultural highlights of the fair evident today include Palm Canyon, just northwest of the Spreckles Organ Pavilion, planted with hundreds of skyline-high palm trees. Also built at this time was the giant lath house, now called the Botanical Building. This striking landmark is located at the northern end of the park. It was used during the Garden Fair to house rare subtropical and tropical plants—how it is used today.

California Pacific International Exposition—A few years after the Garden Fair, hard times for the country and the region slowed the growth of the park area. It wasn't until 1935 that the California Pacific International Exposition opened, with more plants, more buildings and enthusiasm for expansion.

Special Gardens of Today's Balboa Park

Today, the park offers activities and facilities ranging from museums, to theatre, to the world-renowned San Diego Zoo. It is described on page 50.

Balboa Park's Botanical Building, with the Lily Pond in foreground, was built for the 1915-1916 Panama-California Exposition. It is still used today to showcase rare tropical and subtropical plants.

The Alcazar Garden at Balboa Park was inspired by gardens surrounding Alcazar Castle in Seville, Spain.

Special gardens to visit include the Inez Grant Parker Memorial Rose Garden, which includes an All-America Rose Selection garden. It is located south of the footbridge that crosses Park Boulevard near the Natural History Museum. The Desert Garden is located north of the footbridge near the Natural History Museum. It features many species of cacti and other plants indigenous to desert regions of the world. The Alcazar Garden was constructed for the 1915 exposition, designed after Alcazar Castle in Saville, Spain. Precisely trimmed, low boxwood hedges surround spacious squares and rectangles planted with bright displays of flowering annuals.

An elaborate Japanese Friendship Garden will be in operation in the future, offering another reason to visit this multi-faceted city park. The master plan also calls for reduced vehicular traffic in the park's interior, designed to open up space to develop even more gardens and plazas.

Bougainvillea

The San Diego Zoo
PO Box 551
San Diego, California 92112
(619) 234-3153 or 231-1515 for
recorded information.
(619) 231-0251 business office.

Located in the northwest corner of Balboa Park. From I-5 take I-8 east to Highway 163 south. Take the Park Blvd. exit and follow signs.

The San Diego Zoo is one of the finest zoological gardens in the world. Over 6,500 plant species and approximately 3,500 animals can be seen on 100 landscaped acres.

Open daily Post-Labor Day through June from 9 to 4 p.m. From July through Labor Day open 9 to 5 p.m.

Entrance fee required. Children under 2 years free.

Guided bus tours available for fee. Skyfari aerial tram, children's zoo and shows included with entrance fee.

Many gift shops; hours as the zoo.

Restaurant hours: as the zoo.

Video and still camera rentals available.

Wheelchair access.

Special group facilities available for clubs, organizations, companies and parties. Contact business office.

Free animal shows (days and times vary) include Channel Island Sea Lion Show, Animals in Action and Animal Chit Chat Show.

San Diego Zoo

Zoos are no longer entertainment facilities—places to view unusual or exotic animals. They have become invaluable preserves—for animals as well as plants. The San Diego Zoo, long known for its quality animal exhibits, is also an extensive, complex, botanical garden—a member of the American Association of Botanical Gardens and Arboreta (AABGA). More than 6,500 plant species grow on the grounds, many both rare and valuable.

Along with their protective role, the look of zoos is undergoing a metamorphosis, with San Diego at the forefront. The traditional method of grouping animals together by class, such as all primates, or all the big cats, often in cages, is being replaced by organizing animals and plants according to *bioclimatic zones*. Naturalized enclosures, landscaped with the plants indigenous to the particular environment, will create new, more interesting, more comfortable homes for the zoo's inhabitants. In some instances, plants—both native and introduced—will provide food and shelter for the animals.

Tiger River: A Look into the Future

A prime example of the new look in zoos is San Diego's Tiger River, a 3-acre model of an Asian rain forest. (Another is the Sun Bear Forest, a recent addition.) A walk along its 1,500-foot path will take you past more than 100 animals and 500 plant species—more than 9,000 plants total. Many plants here are rare. Those worth noting include the large kentia palms at the entrance, and a China lace tree near the first bench heading down the trail. Around the bend is a Syke's coral tree. Further along near the upper cliff face is a Saigon orchid vine and large, angel-wing tree begonia. Bamboos are also planted throughout the exhibit; look for *Bambusa vulgaris* 'Vittata;' it has distinctive, vertical green stripes.

Technology allows rain forest plants to survive in San Diego's low rainfall (less than 10 inches annually). A fogging system, with 300 high-pressure, mist emitters, creates the cool, misty climate the plants require. These high-pressure emitters operate on timers for a few minutes about every half-hour—using only about 15 gallons of water a day. The effect, shown in the photo, is pleasing for plants, animals and visitors.

Ten animal exhibits are located along the Tiger River route, positioned behind open moats, glass and piano wire. Because of the terrain and jungle-like growth you sometimes have to look closely for the endangered Sumatran tigers, fishing cats, water dragons and Burmese pythons, to name a few.

San Diego's climate allows a wide, wide range of plants to grow and thrive on the grounds. Plant collections include Palms, Cycads, Bromeliads, Orchids, Erythrina, Ginger, Euphorbia, Bamboo and Aloe. One outstanding exhibit is Fern Canyon. (See photo, page 8.) Visitors gain a real sense of a tropical jungle—layer upon layer of plants jockey for position to gather the sun's rays. In the nearby Rain Forest Aviary, birds literally whiz by as you make your way up and down the series of switchback paths.

The 30-year master plan for the San Diego Zoo will bring many changes and improvements, as more exhibits are converted into bioclimatic zones similar to Tiger River. Eventually, The zoo will be reorganized into 10 climate zones: *Rain Forest, Seasonal Tropic Forest, Savanna, Temperate Forest, Grassland, Montane, Taiga, Tundra, Desert* and *Island*.

San Diego Wild Animal Park

This wildlife preserve of the zoo is located 30 miles to the north. Large groups of animals such as rhinos, giraffes and antelope roam over the 1,800 acres, 600 which are landscaped. Over 1.5 million plants and 3,500 species are on the grounds. The park has many specialty gardens to enjoy, including the Nicholas T. Mirov Conifer Arboretum. To arrange a tour of any of the specialty gardens, call (619) 747-8702.

Sumatran tiger

Mist emitters operate for a few minutes every half-hour to create the moist, humid climate required by rain forest plants at the zoo's Tiger River exhibit.

A yellow-beaked milky stork poses for visitors in the rain forest aviary.

Naiman Tech Center
9605 Scranton Road
San Diego, California 92121
(619) 453-9550

From the I-805 Freeway, take Mira Mesa Blvd. east to Scranton Road, go up the hill to parking area near the large red sculpture.

5.5 acres of Japanese gardens, including pond, rock garden and authentic Japanese teahouse and restaurant, integrated into an office and recreation complex.

Open Monday-Friday from dawn to dusk. Closed weekends and major holidays.

For group tours, contact property manager in advance.

No entrance fee.

Restaurant hours: 6:30 to 3:30 p.m. Sushi bar and deli restaurant for breakfast and lunch.

No picnic facilities.

No wheelchair access.

Facilities available for weddings and special occasions at the Kitayama restaurant. Call 457-1444 for information.

The Naiman Tech Center Japanese garden is part of a private office complex, but visitors are welcome.

Koi fish

The Naiman Tech Center Japanese Garden

This exquisite Japanese garden is actually an integral part of a high-tech office complex—the Naiman Tech Center in San Diego. This progressive facility was featured in a *Newsweek* magazine article, "Offices of the Future." The complex could be described as a marriage of East and West, with gleaming office buildings rising eight stories above 5½ acres of gardens, pond and Japanese pavilion. These serve as a central court and visual relief for office occupants and visitors.

The Kitayama pavilion is an authentic reproduction of a 16th century Japanese tea house. Its authenticity was ensured by importing Japanese craftsman for its construction. Inside are a sushi bar and private Tatami rooms for lunches, and a deli-restaurant open for breakfast and lunch. The outdoor patio of the restaurant overlooks the pond, a rock garden and waterfall. Numerous koi fish swim in the pond, surrounded by tastefully sculptured Japanese black pine and bright green mounds of pittosporum. Tight rows of bamboo in the background serve as vertical screens. (Although the restaurant and park facilities were built for the office tennants, visitors are welcome.)

The garden, a narrow rectangle, emulates a natural mountainscape with numerous mounds and hills. Not only do the hills and valleys add interest, they also screen the volleyball, tennis courts, pool and workout facilities that are part of the complex. A half-mile jogging trail is actually part of the grounds, blending in with the gardens.

Alice Keck Park Memorial Garden

Alice Keck Park is in the center of Santa Barbara. It is actually a series of gardens situated within a 4½-acre city block, with seven entrances from the four streets that surround the park. The theme of Alice Keck Park is well-integrated groupings of trees and shrubs, rather than collections of plants. It is a garden that provides color and interest for each season. Plantings of flowering bulbs provide seasonal displays, while under-plantings of perennials and shrubs produce masses of seasonal blooms when bulbs are out of bloom. It is also a delightful place to relax.

Historical Background

The gardens were completed in 1979 and given to the City of Santa Barbara by long-time resident Alice Keck. Landscape architect Grant Castleberg developed the design and landscaping around the many original trees and plants.

Special Features

A giant, elevated sundial is positioned in the center of the garden. Nearby is a group of tall, yellow-flowering sweet shade, *Hymenosporum flavum*. An information panel in this area identifies a majority of the garden's plants. The garden is home to over 100 species of unusual flowering and shade trees, plus a large number of palms adapted to the coastal climate. The result is a colorful and subtropical setting.

Spectacular trees include purple orchid tree, *Bauhinia variegata*, and evergreen Hong Kong orchid tree, *B. blakeana*. Red-flowering gum, *Eucalyptus ficifolia*, a collection of coast-adapted coral trees, *Erythrina* species, lavender-blue flowering jacaranda, *Jacaranda mimosifolia*, and four species of yellow, winter- and spring-flowering *Cassia*. Date palm, *Phoenix canariensis*, and tall Mexican fan palm, *Washingtonia robusta*, add skyline presence. Many of the trees now growing in Southern California and desert area gardens were originally tested by enterprising nurserymen in Santa Barbara during the early 1900s.

A large, rock-edged pond at the center of the garden reflects the skyline palms and mature, canopy-shaped trees. Large koi fish swim among water lilies and other aquatic plants. When combined with the sound of water from gently flowing streams, the mood is delightfully tranquil. A small gazebo at the pond's edge offers a vantage point to overlook the garden and city, while paths and changes in levels throughout the garden invite exploration.

Alice Keck Park Memorial Garden
1500 Block Santa Barbara Street
Santa Barbara, California 93102
(805) 564-5433

Mailing address:
City of Santa Barbara
Parks Department
P. O. Box 1990
Santa Barbara, CA 93102

———

Park is bordered by Alameda, Garden, Santa Barbara and Micheltorena Streets. From US 101, drive 3.5 miles to Garden Street.

———

The park covers a 4-½-acre city block. A delightful place overflowing with natural beauty.

———

Open daily—no formal entry gates. Closes each day at 10 p.m.

———

No entrance fee.

———

Picnic and restroom facilities at nearby Alameda Street Park.

———

Wheelchair access.

———

Park is available for weddings and private gatherings. Contact the Parks Recreation Department
(805) 564-5418.

———

In the 1890s, horticulturist Dr. Francesco Franceschi introduced plants from the world's arid climates to Santa Barbara. During 1920 to 1930, E. O. Orpet introduced appropriate plants for the Santa Barbara area—those requiring little water. Their contributions to many of the park and street plantings have made Santa Barbara a horticultural landmark.

Bottle tree

Santa Barbara Botanic Garden
1212 Mission Canyon Road
Santa Barbara, California 93105
(805) 563-2521 (recorded information)
(805) 682-4726 (business)

From south Highway 101, turn west on Foothill Road. Turn north on Mission Canyon Road to the garden entrance. From north Highway 101, turn east on Mission Street to Laguna Street, turn north to Los Olivios, and proceed past Santa Barbara Mission to Foothill Road (Hwy. 192) to Mission Canyon Road.

Santa Barbara Botanic Garden is a serene garden, where visitors can find natural beauty and scholastic enterprise side by side.

Open daily 8 a.m. to sunset.

Entrance fee required. Free admission every Tuesday and Wednesday.

No picnic facilities.

Wheelchair access to street level garden.

Gift shop hours: 10 to 4 p.m. daily. 10 to 4:30 p.m. April-Labor Day.

Retail nursery hours: 10 to 3 p.m. on Tuesday, Thursday, Friday and Saturday and 11 to 3 p.m. on Sunday.

Library hours: By appointment 9 to 5 p.m. weekdays.

Annual events include: *Give the Earth a Hand Day*, end of September; *Spring Plant Sale*, March; *Fall Plant Sale*, October; *Family/Members Day*, June.

Santa Barbara Botanic Garden

A drive up scenic Mission Canyon Road at the base of the Santa Ynez Mountains to the Santa Barbara Botanic Garden foreshadows the variety and complexity of Southern California's plant population. The garden entrance is especially picturesque, with a view across the colorful meadow and the immense, Blaksley Memorial Boulder, surrounded by a backdrop of mature, California live oaks, *Quercus agrifolia*.

A California Native Preserve

The 65-acre botanic garden is a display and preserve of California native plant species, grouped by the geographical regions of the state. It is fascinating to stroll among northern California redwoods, then be among cacti native to Southern California deserts a few moments later. In addition, the limited water resources of Santa Barbara region has made the garden a valuable source of information on low-water landscape plants and new, water-efficient irrigation methods.

Despite the large variety of plant species on display, there are a few select plants that best represent the garden's nature and spirit: California sycamore, *Platanus racemosa*, California lilacs, *Ceanothus* species, islands ironwood, *Lyonothamnus floribundus asplenifolius*, California coastal redwood, *Sequoia sempervirens*, and California poppy, *Eschscholzia californica*. Look for these in sunny areas throughout the gardens.

Trail System

Over 10 designated trails and pathways wrap around the diverse gardens, covering some 5.5 miles, from meadows to steep canyon slopes. The map provided at the Admissions Center will help you choose which sections to visit, or follow the trail loop signs suggested. As you walk along these paths and throughout the garden, be aware there are over 1,000 species of rare and indigenous California plants on the grounds.

Botanical Research

On a clear day, you can see the Channel Islands in the Pacific Ocean. The islands, located 30 miles from the coast, have been the subject of horticultural research on revegetation techniques, fire ecology, impact of livestock grazing on vegetation and distribution of rare plants. Recent studies include the propagation of native plants and a breeding program for plants with home landscape value.

The Garden Library

The library collection includes 7,700 books and 2,000 volumes of bound periodicals, and has become a valuable research tool for botanists, students and horticulturists. It is located just behind the shop adjacent to the reception office. A herbarium, a climate-controlled archive, stores over 90,000 seed and plant specimens collected over the past 60 years. It is available for use by appointment to qualified participants.

A home demonstration garden is scheduled for completion in 1991. It will provide visitors with valuable examples of how to successfully create and care for a home garden composed solely of drought-tolerant, fire-resistant, Southern California native plants.

A tour group pauses to examine the profuse blooms of matilija poppy, Romneya coulteri.

California sycamore

**Wrigley Memorial and
Botanical Garden
1400 Avalon Canyon Road
Santa Catalina Island
Avalon, California 90704
(213) 510-2288**

———

Located one mile from the City of
Avalon on Catalina Island, 26 miles
west of Long Beach.

———

Approximately 3 acres of developed
gardens on 35 acres. Includes cacti
and succulents, and plants native to
the California islands.

———

Open daily 8 to 5 p.m.

———

Entrance fee required; children under
12 free.

———

Wheelchair access.

———

No picnic facilities.

———

Facilities for weddings available by
prior arrangement.

———

Wrigley Memorial and Botanic Garden

Santa Catalina Island is 26 miles by boat or plane from the mainland out of San Pedro Harbor or Long Beach. The resort town of Avalon is home to Wrigley Memorial and Botanic Garden, located a mile above the resort at the head of picturesque Avalon Canyon. The 3 acres of developed garden is surrounded by 35 acres of native vegetation along the canyon slopes.

The garden is a pleasant and relaxing place, blessed with an almost-ideal climate for plant growth. Not only is it possible to grow plants here that cannot be grown on the California mainland, Wrigley is host to as wide a variety of plants as any location in the world.

Historical Background

William Wrigley, Jr., who founded the chewing gum that carries his name, bought control of the Santa Catalina Company in 1919. Until his death in 1933, he poured his energy and enthusiasm into improving the island. New steamships, public utilities, hotel, the landmark Casino, plus extensive plantings of trees, shrubs and flowers throughout the island, were among the many improvements.

In 1934, Mrs. Wrigley continued to develop the garden. In 1969, the Memorial Garden Foundation re-dedicated the garden with the purpose of creating a living laboratory for studying plants of special interest. Emphasis has been placed on *endemic plants*—native only to Santa Catalina Island. If a plant is categorized as an *insular endemic,* it is a plant that grows naturally on one or more of the California islands but does not grow naturally on the mainland. Most of the plants growing on Catalina Island are water-efficient and adapted to grow under difficult conditions.

The Special Catalina Island Plants

If you are a horticulturist or home gardener in coastal Southern California, you may be familiar with some of the plants described here as endemics. However, some are more common and easier to grow at home in the native island environment. A few botanic gardens, including Santa Barbara Botanic Garden and Rancho Santa Ana Botanic Garden in Claremont, have representative plantings from Catalina Island in their research programs. Plants native to Catalina Island, and found throughout the gardens include: Catalina ironwood, *Lyonothamnus floribundus floribundus;* Catalina manzanita, *Arctostaphylos catalinae;* toyon (also California holly), *Heteromeles arbutifolia macrocarpa;* Catalina cherry, *Prunus lyonii;* lemonade berry, *Rhus integrifolia* and Catalina currant, *Ribes viburnifolium.* They can be found in the wild in only one other location— Saints Bay in Baja California.

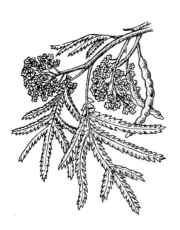

Catalina ironwood

Los Angeles State and County Arboretum

An artistic blending of beautiful landscape scenes with practical demonstration gardens help create the ambience of Los Angeles State & County Arboretum. Add in a boisterous flock of peacocks, an 1880s Queen Anne Cottage recognized for appearances in films and television, and extensive, world-wide plant collections, and it's simple to understand why the arboretum is host to over 500,000 visitors each year.

Plants and Plant Collections

Currently, there are over 30,000 permanent landscape plants on the grounds. Most are grouped according to geographical locations around the world, with emphasis given to regions with climates similar to those in Southern California. Visitors can walk among individual sections of plants native to *Australia, the Mediterranean, North-American-Asiatic, South Africa* and *South America.* Current major collections of plants within many of these gardens include: *Acacia, Callistemon, Cassia, Cycad, Erythrina, Eucalyptus, Ficus, Magnolia, Melaleuca, Orchids* and *Tabebuia.*

One of the more practical aspects of the arboretum is the testing of plants from similar climates for potential use in Southern California. Since 1948, (when the state of California acquired the land for the arboretum), more than 100 flowering plants tested successfully at the arboretum are now used in Southern California landscapes. Some of the best-known include the purple and white freeway daisies, *Osteospermum fruticosum,* yellow *Cassia* trees and pink floss-silk tree, *Chorisia speciosa.*

The Queen Anne Cottage and the Lasca Lagoon. The cottage was built in 1885 by Lucky Baldwin.

Los Angeles State & County Arboretum
301 North Baldwin Ave.
Arcadia, California 91007-2697
(818) 446-8251 or (213) 681-8411
———
Located 20 miles east of downtown Los Angeles. From the 210 Freeway, take the Baldwin Ave. exit south about 1 mile to entrance.
———
Over 127 acres of landscaped grounds, demonstration gardens, plant collections and historic buildings.
———
Open daily 9 to 4:30 p.m. Closed Christmas Day.
———
Entrance fee required.
———
Narrated tram tour available for a nominal fee. Free guided tours of selected historical buildings; call for information.
———
Gift shop hours:
as the arboretum hours.
———
Restaurant hours:
as the arboretum hours.
———
Picnic facilities at arboretum entrance.
———
Wheelchair access.
———
Facilities available for weddings, parties or other special events by prior arrangement.
———
Highlights of special annual events include *Bonsai Show,* last weekend in January, *Valentine Plant Sale,* orchids & tropical plants, early to mid-February; *Environmental Education Fair,* early to mid-March; *Insect Fair,* mid-March; *Baldwin Bonanza Plant Sale,* May; *Queen Anne Cottage & Santa Anita Depot Open House,* December.

The ground cover display at the arboretum gives area homeowners a chance to compare plant color, size and texture.

Freeway daisy

Historic Buildings

The arboretum was once part of the 46,000-acre Baldwin Ranch, owned by Elias J. "Lucky" Baldwin. Baldwin built the Queen Anne Cottage in 1885, and the Coach Barn in 1879, which housed many of Baldwin's favorite carriages. Both buildings were restored in the 1950s by the California Arboretum Foundation. Other buildings include the Hugo Reid Adobe, built by Reid in 1839. Reid was the first private owner of the 13,319-acre Rancho Santa Anita. The building was recontructed in 1957, then later restored to its current state in 1984. The old Santa Anita Depot, constructed in 1890 by the Santa Fe Railroad, was moved due to freeway construction and rebuilt at the arboretum in 1970. It is filled with railroad memorabilia and artifacts from the late 1800s. Free tours of the building are provided.

Some Special Gardens

Several gardens and greenhouses are located around the arboretum grounds. Some to place on your visit-list include:

Home Demonstration Gardens—This series of gardens and landscapes is located to the right just as you pass through the entrance gates. The gardens are actually four backyard landscapes, each one suited to fit a particular lifestyle. Here are designs for entertaining, for a family, for a plant collector and for a low-maintenance gardener. Nearby, to the left at the entrance, is the ground cover display. Here you can see and compare, side by side, the best ground covers for Southern California landscapes.

Gardens for All Seasons—Continuing north from the entrance, beyond the Ayres Hall building, you'll find a conglomeration of mini-gardens: *Gardens for All Seasons*. Planted and maintained solely by volunteers, these gardens present creative ways to combine and grow plants. (The area is fenced to keep out the wandering peacocks.)

Greenhouses—The Begonia Greenhouse, Tropical Greenhouse, and Orchid Greenhouse are located just north of the Gardens for All Seasons. You'll first come upon the Begonia Greenhouse, home to some 200 varieties. A bit farther north is the Tropical Greenhouse, and within it, the Orchid Greenhouse. The Orchid Greenhouse contains one of the largest collections in the United States—over 10,000 flowering species and hybrids.

Henry C. Soto Garden of Drought-Tolerant Plants—Located just west of the Gardens for All Seasons, this garden is dedicated to the former president of the California Landscape Contractors Association. It features dozens of species of low-water use plants, located in six planting beds. Visitors are exposed to new landscape plants and plant combinations that are water-thrifty alternatives to traditional high water-use plants.

Lasca Lagoon—If this scene looks familiar, you might have watched a television show popular in the early 1980s, *Fantasy Island*. The lagoon with the Queen Anne Cottage in the background were featured in the show's introduction. The lagoon is actually a natural lake that covers 3½ acres. Early Indian artifacts dating back to pre-Columbian times have been discovered around its shores.

Prehistoric & Jungle Collection—Walking south from the entrance toward Lasca Lagoon and the Queen Anne Cottage will take you to the

Prehistoric & Jungle Collection. It, too, has been featured in many films, including *Tarzan* and *The Road to Singapore.* The cycad collection is located here, as well as an Interpretive Center that explains the evolution of plants.

Meadowbrook Area—Located west of the historic buildings and Lasca Lagoon, The Meadowbrook covers about seven acres. It is a re-creation of a temperate zone meadow, complete with a meandering stream. The Grace Kallam Perennial Garden is filled with hundreds of flowering perennials. Notice, too, the rock gardens located along its western edge.

Herb Garden—This garden is located beyond the historical buildings, southeast of Meadowbrook. It covers 1½ acres and contains more than 450 herbs, grouped according to how they're used. Included within is the Braille Terrace. A waist-high wall guides visitors to touchable, fragrant herbs, each of which are identified in English and Braille. Also included here is the Shakespeare Garden, where plants mentioned in his works are arranged in formal displays.

Waterfall & Aquatic Gardens—These gardens are located just south of the Meadowbrook area. Visit these gardens if you're looking for shade and relaxation. A man-made recirculating waterfall cascades about 50 feet into a pool inhabited with colorful koi fish.

The vivid yellow flowers of the trumpet tree, Tabebuia chrysotricha, create a brilliant springtime show at the arboretum.

Rancho Santa Ana Botanic Gardens
1500 North College Avenue
Claremont, California 91711
(714) 625-8767

Located 30 miles east of Los Angeles.
From I-10 take Indian Hill Boulevard
exit north to Foothill Boulevard. Travel
east to College Avenue. Go north on
North College Avenue to
the garden entrance.

85 acres of native California plants,
including more than 1,500 species.

Open daily 8 to 5 p.m.

Closed July 4th, Thanksgiving Day,
Christmas Day and New Year's Day.

Tours at 2 p.m. during weekends,
March-May, or by arrangement.

No entrance fee.

Bookstore hours: 8 to 5 p.m.
weekdays; 11 to 4 p.m. on weekends.

Library hours: 8 to 5 p.m. weekdays,
by appointment only.

No picnic facilities.

Wheelchair access.

Restaurant facilities planned for the
future.

Wedding and receptions allowed in
selected areas. Call for information
and reservation.

Annual events include:
*Southwestern Systematics
Symposium,* May; *Ecological
Landscape Symposium,* October; and
Plant Sale, first
Saturday of November.

Rancho Santa Ana Botanic Gardens

A visit to Rancho Santa Ana Botanic Gardens takes you into a living museum of California natural history. Over 1,500 species of plants are on site, including the rich plant heritage of upper Baja California. More than one-fourth of all known species of native North American plants, excluding Mexico, can be seen on the grounds. It is interesting to note that California has the most diverse assemblage of plants of all the continental states.

The gardens cover 85 acres of landscaped areas, with some situated on a mesa known as "Indian Hill;" others are located at lower elevations. The size of the garden and variety of terrain provide a unique opportunity to see and learn about a wide selection of California's native plant communities.

Representative Plants

The spirit of the garden is represented by California natives such as California lilac, *Ceanothus* species; and Western redbud, *Cercis occidentalis*. California poppy, *Eschscholzia californica*, and coral bells, *Heuchera* 'Santa Ana Cardinal', are well-distributed throughout the grounds. In addition, the popular Leyland cypress, *Fremonta* 'California Glory', *Ceanothus* 'Frosty Blue', and *Mahonia* 'Golden Abundance', are Rancho Santa Ana introductions. Stream-side plantings with drifts of naturalized iris and wildflowers, shaded with mature sycamore trees, add to Santa Ana's quiet, reflective atmosphere. Other areas of the grounds provide ideas and examples of rock gardens, and ways to use plants in arid regions.

Began As A Spanish Rancho

Rancho Santa Ana was founded in 1927 by Susanna Bixby Bryant in memory of her father, John W. Bixby. The site was a parcel of a Spanish rancho in Santa Ana Canyon in Orange County. In 1951, the facilities were moved to the current location in Claremont. The nearby Pomona College and the Claremont Graduate School would later become affiliated with the gardens for botanical research. The garden serves as the graduate program in botany for the graduate school—the only botanic garden in the country to offer such a program.

The meandering paths that crisscross Rancho Santa Ana are bordered by native plants in *desert*, *chaparral*, *woodland* and *forest* settings. Each provides an opportunity to explore, in a natural state, the native landscapes of California. Visiting the garden areas at different times of the year allows you to catch the beauty of each season. For the most colorful displays, the wildflowers are in bloom March through mid-May.

Research and Education

Finding new ways to conserve water and methods of re-establishing endangered plants are important goals. Staff members share and communicate information culled from over 60 years of research on these vital issues. Information and ideas on how to use natives and drought-tolerant plants introduced from similar dry climates educate professionals and home gardeners alike. In addition, botanists such as Dr. Philip A. Munz and Dr. Lee W. Lenz have developed valuable plant hybridization programs, and authored numerous volumes on the California flora at Rancho Santa Ana.

For the botany student, the gardens house a comprehensive herbarium that contains over one million specimens. A library of horticulture and botany references is on-site, plus a seed storage house for endangered species and exchange programs.

Demonstration Garden

The delightful home demonstration garden is located northwest of the Plant Science Center, laboratories, offices and bookstore. Raised planters and architectural features help showcase California native plants. This garden is also the site of numerous weddings and parties.

In the Future

A master plan for Rancho Santa Ana includes a garden of California cultivars; an ethnobotanic display; permanent wildflower areas; and "A Walk Through California Trail." A visitor's orientation center and gift shop, restaurant facility and amphitheater will also be constructed to serve the ever-increasing number of visitors.

Coral bells

White-flowering coral bells, Heuchera 'Santa Ana', serves as a perfect understory plant beneath a California live oak, Quercus agrifolia, at Rancho Santa Ana.

Fullerton Arboretum

Fullerton Arboretum
California State University, Fullerton
P. O. Box 34080
Fullerton, California 92634-9480
(714) 773-3579

Located at the northeast corner of the University campus between Yorba Linda Boulevard and Nutwood. From the 57 Freeway, take Yorba Linda Boulevard west to Associated Road, then south to entrance.

Open daily 8 to 4:45 p.m. Closed Thanksgiving, Christmas and New Year's Day.

No entrance fee.

Heritage House, an important historical part of the arboretum, is furnished with vintage Victorian furniture and gardens to match. Hours are 2 to 4 p.m. Sunday. Weekday tours available by reservation one month in advance. Nominal admission fee.

Gift and garden shop hours: 11 to 2 p.m. Tuesday to Saturday; 1 to 4 p.m. Sunday. Closed entire month of August.

Library hours: Open to the public for research Thursday 12 to 3 p.m and Saturday 1 to 4 p.m. Closed during August.

Picnic facilities.

Wheelchair access.

Workshops, demonstrations, displays and seminars available throughout the year. Call for information.

The Heritage House, built in 1894, was the home of Fullerton's first doctor, Dr. George C. Clark. Plants around the home represent ornamentals commonly grown at the turn of the century.

This 26-acre green oasis located in the midst of Orange County's urban sprawl could be described as a "garden for all seasons." Botanical collections from around the world form the framework of the garden. Those who visit Fullerton Arboretum tend to return to discover and learn more about plants, as well as to relax and enjoy its natural beauty.

The arboretum provides the community and university with plant and gardening education, conservation and research. In 1970, the idea for establishing the arboretum was fostered by university students, faculty and individuals in the community. Within two years, 26 acres were set aside; after nine years of planting and development, the gardens officially opened October, 1979.

A Walking Tour

The "Self-Guided Tour" brochure available at the entry will lead you through the garden grounds. As you enter the garden, a waterfall and small wooden bridge mark the entrance. Just above and beyond the waterfall, a slope is lush with colorful, drought-tolerant plants adapted to

Southern California. A bit farther down the path, temperate zone plants become numerous, with groves of conifers—from pines to redwoods.

If you follow the well-planned tour, you'll be exposed to the three major plant groupings at the arboretum: *Temperate Zone, Tropical Zone* and *Arid Zone.* Strolling the paths, notice how plants from regions around the world are grouped according to moisture requirements. The stream bed and pond environments also provide a sanctuary for a large population of birds, fish and pond turtles.

The Tropical Zone, located at the southern end of the grounds, features cold-tender plants. These include several *Ficus* species, coral trees, *Erythrina* species, *Calliandra* and even a collection of bromeliads. The Arid Zone includes a palm collection, Australian section, Desert Woodland and plants from other dry climates of the world.

The diverse plants of South Africa and South America blend in with other plants on the grounds. California chaparral and foothill natives are also well-represented, as are plants indigenous to coastal sage scrub. Plant collections from the California Channel Islands are also included. These plants include the Catalina ironwood *Lyonothamnus floribundus asplenifolius,* lemonade berry, *Rhus integrifolia,* and Catalina cherry, *Prunus lyoni.* Completing the Arid Zone collection is an area devoted to oaks, *Quercus* species, set in a grassy meadow.

A deciduous fruit orchard, dwarf citrus, community garden plots (available on a first-come basis) and the Rose Garden provide a transition to the Heritage House entry gardens. Seasonal annuals are arranged here in traditional displays.

Drought-tolerant slope planting located just inside the arboretum's entry is a collection of colorful ground covers suited to warm, dry climates. Plants include the red-flowering Galvesia speciosa, native to Catalina Island, and the pink, Mexican evening primrose, Oenethera berlandieri.

Silk oak

Descanso Gardens
1418 Descanso Drive
La Canada, California 91011
(818) 952-4400

From the 210 Freeway, take Foothill exit south on Angeles Crest Highway, west on Foothill Blvd. to Verdugo Blvd; go west to Descanso Drive, then south to garden entrance.

Over 165 acres, including a 30-acre camellia forest containing over 100,000 plants.

Open daily 9 to 5 p.m. Closed Christmas Day.

Entrance fee required.

Narrated tram tour available for a nominal fee. Special group tram tours available.

Gift shop hours: as the garden hours.

Restaurant hours:
Cafe Court 10 to 3 p.m.
Japanese Tea House open 11 to 4 p.m.; closed Mondays.

Picnic facilities adjacent to entrance parking lot.

Wheelchair access.

Facilities available for weddings, parties or other special events by prior arrangement.

Highlights of special annual events include *Spring Garden Show* during April; *Performing Arts Programs Under the Oaks*, Sunday afternoons May-August; *Arts & Crafts Faire*, June; *Plant Sale*, October; *Christmas Show* during early December.

Camellia

Descanso Gardens

Like many public gardens, Descanso Gardens began as a private residence. The first to develop the property was E. Manchester Boddy, the owner and editor of the Los Angeles *Daily News*. Boddy named the property Rancho del Descanso, *Ranch of Rest*. His home, the Hospitality House, built in the 1930s, overlooks the garden grounds. It is used today to house monthly art show exhibits.

Descanso is one of four gardens operated by the Los Angeles County Department of Arboreta and Botanic Gardens. The others are Los Angeles State and County Arboretum, described on pages 57 to 59; South Coast Botanical Garden, pages 46 to 47; and Virginia Robinson Gardens, page 81.

Special Gardens

Descanso is a camellia-lover's paradise. In fact, the camellia collection is the largest in the world. Over 100,000 plants—some the size of substantial trees—share 30 acres with a forest of native California oaks, *Quercus agrifolia*. Numerous trails wind in and around the rolling hills and among the camellia groves, evoking a feeling of quiet seclusion and relaxation.

Rose aficionados will not be disappointed, with two gardens to visit. The garden History of the Rose features over 200 historical varieties. Covering over five acres, paths guide visitors to roses that were favorites of the ancient Romans. Others include varieties popular through later historical time periods up until the present. The Modern Rose Garden is filled with over 3,000 roses. Almost every winner of the American Rose Society's All-America Rose Selection Award (since 1939) is on display.

Rugged, chaparral-covered slopes make up portions of Descanso's grounds. The extensive California Native Plant Garden, located in the eastern section, covers 15 acres. In spring months, wildflower plantings can be very colorful. Here you can also see flowering perennials, conifers, oaks, ceanothus and maple. Many of these plants are naturally drought-tolerant, and will prove to be good landscape choices as water becomes a more precious resource.

Oriental Pavilion

The Tea House and Japanese Garden is a serene, shaded location. It's a perfect place to pause, reflect and enjoy beverages and cookies served by kimono-clad waitresses on the patio overlooking pools, stream and Japanese bridge.

Color by the Seasons

Flowers are in bloom year-around at Descanso. As the camellia blooms fade, azaleas and iris pick up the pace, followed by roses and native plants. Tens of thousands of bedding plants and bulbs, including 15,000 daffodils, put on quite a show. Here's a monthly account of what you can

Orange and yellow Oriental poppies surround a sundial. Large red-flowering tree in background is a camellia—one of 100,000 camellia plants on site.

A traditional Japanese bridge at Descanso; red and pink azaleas add highlights in the foreground. Nearby is the Tea House, where snacks and tea are served by Kimono-clad waitresses.

expect by way of garden color:

January: camellias, winter annuals

February, March: camellias, deciduous flowering trees and shrubs, early bulbs, lilacs

April: outdoor orchids, azaleas, iris, spring bulbs, native plants

May: outdoor orchids, roses, ceanothus, annuals, native plants

June: old-fashioned roses, summer annuals

July, August, September: roses, summer annuals

October: sasanqua camellias, annuals, roses

November, December: camellias, berried plants, roses

Lummis Home State Historic Monument
200 East Avenue 43
Los Angeles, California 90031
(213) 222-0546

———

Travel north from Los Angeles Civic Center on the Pasadena Freeway. Take Avenue 43 turnoff, go west to Carlotta Boulevard—the first street west of freeway. Turn south on Carlotta Boulevard; street-side parking only by the entrance.

———

The Lummis home and garden, covering 1.8 acres, demonstrates methods of water conservation around an historic home.

———

Open 1 to 4 p.m. Thursday-Sunday.

———

No entrance fee.

———

Wheelchair access.

———

No picnic facilities.

———

Tour time: approximately one hour.

———

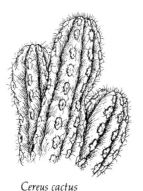

Cereus cactus

The Lummis home and garden provides a fascinating look at a historic Los Angeles home surrounded by a contemporary, colorful, low-water landscape.

Lummis Home
State Historic Monument

This is a garden sensitive to water conservation and relatively low maintenance, presenting practical ways to blend and combine low-water plants. It was recently renovated to present a renewed version of the original landscape, but with new, additional dry-climate plants from around the world. The garden is irrigated just once a week for 15 minutes fall through spring; during summer, watering extends to 30 minutes a week.

The garden's backdrop of mature trees—sycamores, California live oak, and California bay—were planted decades prior to 1928 when the gardens first opened. During spring, flowering plants include native blue lilacs, purple verbenas, wildflowers and rockrose. Crape myrtle and the large, white flowers of matilija poppy put on a show during the summer and fall. *Ceanothus* and the succulent *Senecio* add their shades of blue during spring and summer. A lawn substitute, common yarrow, *Achillea millefolium,* with flat, rose-colored flower clusters, blooms continually.

The Lummis Home was named El Alisal—The Place of the Sycamore—by Charles Lummis. Building the house and landscaping the grounds were a dream-come-true for Lummis, who arrived in California in 1885. He was city editor at the *Los Angeles Times,* and had many friends involved in horticulture, who introduced him to native plants and how to use them. He built much of the house himself over a span of thirteen years, and planted and nurtured the garden until his death in 1928. The arroyostone-covered home features displays of memorabilia, as well as historic photos taken by Lummis.

Landscapes Southern California Style

Landscapes Southern California Style was developed by the Western Municipal Water District of Riverside and the University of California Cooperative Extension at Riverside. It clearly demonstrates how a good design, plant selection and innovative planting and maintenance practices can be used to create an attractive, water-thrifty, low-maintenance garden.

The garden's main emphasis is *education*. It fills the information gap for homeowners, landscape professionals and regional planners, by showing, through step-by-step signage and actual examples, how to create attractive, practical, water-efficient landscapes. Examples demonstrate how water-efficient plantings save money and time in addition to water, while reducing energy needs, producing healthier plants and a more natural-looking landscape. Visitors can learn how to design and plant a new water-efficient garden, and how lawn areas can be replaced with native ground covers to reduce water use. They'll also learn about preparing soil, and how to select and install low water-use drip irrigation systems for shrubs, trees, ground covers and turf. Visitors also become acquainted with colorful dry climate plants, such as *Salvia* species, gazanias, the new Chitalpa tree and *Ceanothus* species.

A Team Effort

Landscape and plant suppliers from Southern California furnished products, services and support to develop the garden. The California Department of Forestry and California Conservation Corps pitched in, providing the labor to install the materials and plants.

Landscapes Southern California Style
450 Alessandro Boulevard
Riverside, California 92517
(714) 780-4170

Located south of Riverside. Take 215 to Alessandro Boulevard, and travel west to the Western Municipal Water District offices and garden.

One-acre garden, thoughtfully designed to educate homeowners and landscape professionals on ways to design, create and care for colorful low-water landscapes.

Open daily 10 to 4 p.m.
Closed major holidays.

No entrance fee.

Call ahead to schedule a guided tour, if desired.

No picnic facilities.

Wheelchair access.

Gift shop hours: as the garden.
Educational pamphlets and seeds only.

Meeting facilities (up to 50 people) related to water conservation, education and environmental quality of landscaping available by reservation.

Special events include: *Water-Efficient Plant Sale,* spring months; *Monthly Garden Walks With Director,* call for reservations; *Water-Conservation Techniques Seminars,* call for schedule.

One of the special features of Landscapes Southern California Style is the thoughtful use of informational signs. Visitors can walk through this demonstration garden and learn, step by step, how to plant and maintain water-efficient landscapes.

University of California Riverside (UCR) Botanic Garden
UCR Campus
Riverside, California 92521
(714) 787-4650

Riverside is 56 miles east of Los Angeles. From I-215/State Highway 60, exit University Avenue or Pennsylvania Avenue and head east. After arriving at the campus entrance, follow Campus Circle Drive to Parking Lot 13. Enter this free parking area to reach the cedar-lined drive to the garden entrance.

Open daily 8 to 5 p.m. Closed July 4th, Thanksgiving, Christmas and New Year's Day,

No entrance fee but contributions gratefully accepted.

Group arrangements for guided tours are available. Call (714) 787-4650. Or write to Curator, UCR Botanic Gardens.

UC Riverside Botanic Gardens

The University of California, Riverside Botanic Gardens (UCR), is an environment of many diverse microclimates. Each *microclimate*—a very small, localized climate, includes collections of plants from various regions of the world. The result is an amazing selection of plant forms, colors and textures, which makes the gardens a pleasure to visit all seasons of the year. In addition, the 4-½ miles of paths are a hiker's delight, providing an opportunity to enjoy the hundreds of bird species, small animals, snakes, lizards and frogs that make the UCR gardens their home.

With the "Outdoor Classroom" self-guided tour (available at the entry), you can explore the garden at your own pace. There are 39 scenic acres of rolling hills, boulder-studded slopes and dry creek beds in the lower canyons, creating natural settings for the more than 3,000 plant species arranged by plant community.

Garden Highlights

A good starting point in your tour is the main east trail that borders the North American gardens of the Mojave and Sonoran Deserts. Here are cacti, succulents, trees and shrubs. Featured plants include blue palo verde, mesquite, Mojave yucca, *Acacia* and jojoba.

As you turn south up the hill, pines and redwoods (conifers) are predominant plants. Adjacent to the conifers is the Rose Garden, with over 200 selections of wild and cultivated forms. Some of the heritage roses, also called *old-fashioned roses,* are quite fragrant and worth sampling. Gardens farther along the trail include perennials, iris and tree peonies. Several members of the cycad family are sheltered within the geodesic dome lath house. Almost opposite the lath house, herbs such as lemongrass, fennel, sage, lemon verbena, rosemary and several kinds of mints are grown.

As you turn north at the bridge and pond, perhaps you'll see koi and bluegill, a common sunfish, or a red-earred turtle or bullfrog. Returning to the entrance gate through Alder Canyon, a variety of trees create a canopy of shade, allowing azaleas, camellias, ferns and winter daphne to thrive.

The Fruit Orchard

The recently established subtropical fruit and deciduous fruit tree orchard has become a center for demonstration and selection of varieties for home and commercial use. Rare fruits from all over the world are being grown commercially, and are available to home gardeners from

Oranges

selected nurseries. (See listings of commercial gardens on pages 86-87.) In the orchard's inventory are kiwis, cherimoya, litchi and macadamia nuts, mango and jujubes, over 30 citrus varieties, plus several varieties of fig, avocado, apples, almonds, cherries, pears and many other fruits.

As the greater Riverside-San Bernardino region continues its rapid expansion, the UCR Botanic Gardens will become an even more valuable horticultural resource—an *outdoor classroom* for visitors and residents.

Flowering plants cover a steep, rugged slope in the South African section at the UCR Botanical Gardens. In the foreground is toadflax, Linaria reticulata. White flowers are cape marigold, Dimorphotheca.

The Huntington Library, Art Collections
and Botanical Gardens
1151 Oxford Road
San Marino, California 91108
(818) 405-2141

From the 210 Freeway, exit at Allen
Avenue and proceed south to the
Huntington's gate at Orlando Road.
From I-10, turn north on Rosemead to
Huntington Drive, drive west
to Oxford Road.

150 acres of exquisite gardens.
Special gardens include *Japanese,
Conifer, Desert, Rose, Shakespeare,
Subtropical, Camellia* and *Azalea,
Jungle, Arbor, Australian* and *Palm.*
More than 14,000 plant species.
Spectacular art galleries and library.

Open Tuesday-Sunday 1 to 4:30 p.m.
Reservations suggested for Sunday
visits. Closed Mondays and
major holidays.

Tours available at 1 p.m. daily.

Donation is suggested.

Gift shop hours: 12:30 to 5 p.m.

Restaurant hours: 1 to 4:30 p.m.

The Irvine Orientation Center adjacent
to the Library provides a publication,
*An Introduction to the Huntington
Botanical Garden.*

No picnic facilities.

Wheelchair access.

Special annual events include: *Plant
Sale,* May; *Friends Day Open House,*
June; *Fall Plant Festival,* November.
Exhibitions in library and art galleries
are changed regularly.

The Huntington Library, Art Collections and Botanical Gardens

Henry E. Huntington, a real estate developer and railroad tycoon, assembled Huntington Gardens from 1903 to 1927. His vision of extensive plant collections and sumptuous gardens resulted in one of the most beautiful gardens in the world. A visit to Huntington is a superb and diverse botanical and artistic journey, one you will surely find memorable.

The history of horticultural development in Southern California has been influenced greatly by the contributions made by Mr. Huntington and those who designed the gardens. Previous to 1928, before the gardens opened, a deed of trust was set up so Huntington's magnificent estate could be maintained in perpetuity. The Huntington was masterfully overseen by botanic garden director William Hertrich, who was also an author of distinction.

"To most horticulturists the Huntington Botanical Gardens and William Hertrich are almost synonymous terms, so closely woven did the two become during the first half of this century. That the Huntington Botanical Gardens is today one of the outstanding repositories of rare plants in the United States is owing entirely to his inspired planning and tireless work over every detail of the garden's development." (From the book, *Southern California Gardens,* by Victoria Padilla.)

The Primary Gardens

Hertrich introduced collections of plants from around the world, and integrated them into a series of gardens. Each was separate in itself, but all contributed to the whole of the Huntington Gardens. As you visit the grounds, each garden seems to blend into another. Rolling lawns, charming vistas bordered by mature canopied oaks and other trees provide the setting for the 15 special gardens on 150 acres.

The Desert Garden—The Desert Garden covers 12 acres and includes 2,500 specimens of cacti, succulents and more fibrous plants. Walking into this garden will make you imagine that you've wandered into another world. You'll come across plants that you never knew existed, in an astonishing array of sizes, shapes, textures and colors. Giant, columnar cacti, multi-stemmed yuccas, agaves, opuntias, cereus, miniature mammillarias, aloes, echinocactus, beaucarneas, euphorbias, crassulas and ocotillo are woven into a world of fascinating, unequalled forms. Virtually all plants in this garden are *xerophytic,* a term applied by botanists to plants structurally adapted to resist drought. Plants are grouped by geographic area, and every season of the year offers spectacular floral displays.

Azalea-Camellia Garden—Just north of the Art Gallery in the tree-shaded North Vista and in the North Canyon are 175 varieties of early- and late-blooming azaleas and some 2,000 camellia cultivars. The most profuse bloom period is in late winter and early spring.

Shakespeare Garden—Just a short walk from the North Vista and Camellia Garden is the colorful Shakespeare Garden. Blooming annuals and perennials commonly used in British gardens take center stage, providing photographers and plant-lovers with colorful vistas year-around. Plantings of pansies, poppies, pink carnations, iris, daffodils, roses, columbine, rosemary and heathers create traditional English garden borders.

Herb Garden—Beyond the Patio Restaurant, you'll find the intricately designed Herb Garden. Herbs are arranged in sections that include plants for medicines; teas, wines and liqueurs; cooking, salads and confections; cosmetics, perfumes and soaps; sachets and insect repellents; and dyes.

Arbor Garden—Moving west from the Herb Garden toward the Japanese Garden, you'll come to the Arbor Garden. Here, the arbor supports colorful climbing roses, fragrant jasmine and trumpet vines.

Jungle Garden—This area includes most of the garden's 50-plus bamboo species. They include clumping and running types, 35- to 40-foot high giants to ground-hugging clumps. The lush growth and structure of

The Desert Garden is one of the Huntington's most extensive gardens, covering 12 acres and featuring more than 2,500 species. Shown are coral-flowering Aloe species.

Ornamental chard adds a bright spot of color in the Herb Garden.

these many species create a jungle-like appearance. Some favorite bamboos are Chinese goddess, *Bambusa glaucescens riviereorum,* and black bamboo, *Phyllostachys nigra.*

Palm Tree Collection—Opposite the parking area is a collection of 200, rare, mature palms from around the world. In the last decade, palm trees have renewed their popularity as landscape elements in both mild climate regions and desert areas. This collection reveals the variety in this spectacular plant group.

Rose Garden—West of the Art Gallery is an extensive, well-maintained rose garden. It includes a collection of more than 1,000 historical rose varieties that are living links with history. Spring and fall are the most profuse flowering seasons. One of the interesting features of many old roses is their intense fragrance—seldom matched by today's roses.

The Shakespeare Garden, with gazebo in the background, features plants common to English gardens during Shakespeare's time.

Japanese Garden—Just west of the Rose Garden is the 5-acre Japanese Garden. Take a soul-soothing visit to the 19th-century Japanese house, moon bridge, temple bell and pavilion surrounded by traditional Japanese plants—pines, magnolias, wisteria, sago palms, plums and bamboo. In the 1960s, the Zen Garden of raked gravel was created to make the Japanese Garden more complete.

Australian Garden—Mature eucalyptus trees, bottle brush, *Callistemon* species, *Acacia* trees and *Melaleuca* are a few of the many species in this garden. These are great examples of the wide assortment of drought-tolerant, Australian plants used in today's southwestern landscapes.

Subtropical Garden—This garden, located on a slope, includes colorful plants from many parts of the world native to *subtropical* climates—the climate zone between the cooler *temperate* and warmer *tropical* climates. Look for yellow cassias, pink cape chestnuts, purple orchid trees, blue jacaranda and red, blue or purple sages. Flowering ground cover plants are also prevalent in this garden.

The Library and Galleries

In addition to its numerous gardens, the Huntington Library features over 2.5 million volumes of historic American and English books and manuscripts. The world-renowned art collection in the Huntington's Gallery contains 18th and 19th century paintings such as the "Blue Boy," rare tapestries, sculptures and furniture. Adjacent to the Shakespeare Garden is the Scott Gallery of American Art. It features paintings and period furnishings from the 1730s to the 1930s.

The gardens conceived and developed by the Huntington Botanical Gardens are a precious resource to residents and visitors to Southern California. Take time to visit if you are in the Los Angeles area.

Azaleas in bloom

The Theodore Payne Foundation

The Theodore Payne Foundation
for Wildflowers and Native Plants, Inc.
10459 Tuxford Street
Sun Valley, California 91352
(818) 768-1802
(818) 768-3533—Wildflower Hotline
from March to May

———

Northwest of Glendale off I-5. Take
the Sunland Boulevard exit north to
Tuxford Street. Turn right on Tuxford
to entrance, which is marked
with a small sign.

———

The Theodore Payne Foundation is a
valuable learning and information
center and nursery for 800 plant
species, including rare and
endangered plants.

———

Open 8 to 4:30 p.m. Tuesday-
Saturday. Summer hours vary: call
ahead. Closed Christmas Day, New
Year's Day, Thanksgiving Day,
Memorial Day and July 4th.

———

No entrance fee.

———

Reservations required for groups of
ten or more.

———

Gift shop hours 8 to 4:30 p.m.

———

Tours by arrangement.

———

No wheelchair access.

———

Library hours 8 to 4:30 p.m.

———

Special annual events include *Open
House and Native Plant Sale*, April;
Summer Clearance Sale, June; *Fall
Plant Sale*, November;
Wildflower Bus Tour, Spring.

———

The Theodore Payne Foundation is an important source of plants and information for Southern California gardeners. Here visitors can see and purchase a wide variety of California natives and wildflowers.

This non-profit organization was founded in 1960 to carry on the work of horticulturist, Theodore Payne. Payne devoted his energies to the preservation and use of California native plants. During the course of more than 55 years, he introduced between 400 and 500 species of wildflowers and native plants, making both seeds and plants available for use. The legacy of Payne's pioneering efforts can now be seen throughout home gardens in Southern California as well as in wildflower preserves such as the Antelope Valley California Poppy Reserve. (See page 82.)

The nursery and gardens encompass 21 acres and contain 800 plant species, including some 100 rare and endangered plants. The 3-acre Wildflower Hill is covered with a beautiful mix of wildflowers from March through May. Landscape gardens of natives and a bird-attracting garden provide striking examples of landscapes to assist gardeners in making their own plant and design choices.

In addition to the extensive plant nursery, the seed room is the source of a large assortment of seeds, including wildflowers, shrubs and trees.

California poppy, Eschscholzia californica, is a trademark of the Theodore Payne garden.

Pageant of Roses Garden

The Pageant of Roses Garden is located at the base of the Puente Hills in a region called The Whittier Narrows. The area's climate is influenced both by the cool coast of Los Angeles and Orange County, as well as the warm interior of the San Gabriel Valley. The growing conditions are near-ideal; roses bloom from March through December.

Entering the garden, you immediately notice the vastness of the plantings—over 7,000 bushes. Among these are history's oldest rose, a moss rose, and the most-current All-America Selection winners. You'll also see test roses that could soon be available, identified only as numbers.

A Garden Tour

The brochure available at the Information Center lists each rose variety and its location. You can also pick up a rose-care pamphlet, "Beautiful Roses," written by local rose experts. The balance of the garden features colorful displays of hybrid teas, grandifloras, hedge roses, tree roses, floribundas and miniatures. Weeping tree roses reach 8 to 10 feet high in cascading blankets of color throughout the garden. Varieties include Margo Koster, China Doll and Renac. Traveling on Workman Mill Road from Whittier or the 605 Freeway, you can also see miles of climbing roses—over 500 of them—bordering the fences at Rose Hills Memorial Park. They represent one of the world's largest collections.

Pageant of Roses Garden,
Rose Hills Memorial Park
3900 South Workman Mill Road
Whittier, California 91748
(213) 699-0921

Located east of Los Angeles. From the 605 Freeway, exit at Rose Hills Road. Head east to Workman Mill Road, then left to the Rose Hills Memorial Park entrance.

Open to the public in 1959. In 1984, the garden received the All-America Rose Selections (AARS) Public Rose Garden Achievement Award, designating it as the most-outstanding public rose garden in the nation. Over 7,000 rose bushes including 600 varieties on 3-½ acres.

Open daily from 8 a.m. to sunset.

No entrance fee.

Flower shop: Open daily 8 to 8:30 p.m.

Library and research center open Monday-Friday 8 to 5 p.m.

No picnic facilities.

Wheelchair access and handicapped parking available.

Annual events include: *Rose Care Seminar*, second Saturday in January; *Mom and Me, Children's Art Festival*, Mother's Day weekend; *Classical Concert Series*, Sundays in September; *Rose Show*, during Veteran's Day weekend.

Roses at the Pageant of Roses garden enjoy a long bloom season–stretching from March to December. These were photographed during April.

The Living Desert
47900 Portola Avenue
Palm Desert, California 92260
(619) 346-5694

Located 15 miles east of Palm Springs. Take Highway 111 to Portola Ave., go south 1-½ miles to entrance.

Over 1,200 acres, with 20 acres of natural desert gardens, live desert animals, trails, demonstration gardens and plant nursery.

Open daily September 1-June 15 from 9 to 5 p.m., including holidays. Closed June 16 to August 31.

Entrance fee required.

Wheelchair access.

Gift shop hours: as the garden.

Picnic facilities available.

Facilities available for private functions, weddings and meetings.

Annual events include: *American Indian Cultural Exhibition,* President's weekend in February; *Navajo Rug Show,* early February; *Masters in Glass,* Thanksgiving weekend; *Spring Plant Sale,* early spring. Call for dates of events.

Bighorn sheep

The Living Desert

The moment you enter The Living Desert, the ecological world of the desert Southwest begins to unfold. The experience becomes one of discovery for children as well as adults. Visitors to the grounds often leave with a better appreciation of the desert, as well as a few of its secrets.

In 1970, an oasis of native California fan palms, *Washingtonia filifera,* and a system of primitive nature trails marked the beginning of The Living Desert. Also at this time, 10 distinct, North American desert habitats were created. Now, more than two decades later, visitors can walk among extensively planted replicas of the Southwest's Mohave, Yuma, Sonoran, Baja, Viscaino and Chihuahuan Deserts.

The Living Desert covers 1,200 acres, 20 acres of which are developed gardens. Over six miles of hiking trails wind to higher scenic elevation vistas, providing an intimate look at the natural desert. The many plant species on site represent the desert's heritage and serve as reminders of the need to preserve this fragile environment.

The Living Desert is the finest example of preserved, protected natural desert in Southern California. Its appeal is enhanced by the wonderful sense of place provided by the majestic 6,000-foot Santa Rosa Mountains to the south, and the views of the Coachella Valley as it stretches to the north and east.

Special Plants of the Gardens

Many desert-adapted plants with interesting or unusual features can be seen on the grounds. For example, you'll see tall California fan palms, *Washingtonia filifera,* desert willow, *Chilopsis linearis,* with its clusters of purple orchid-like flowers, and groves of the distinctive smoke tree, *Dalea spinosa.* Others include red-tipped ocotillo, *Fouquieria splendens,* rugged yet picturesque mesquite, *Prosopis glandulosa,* and blue palo verde, *Cercidium floridum,* which, despite its name, has bright green bark and leaves. Collectively, they serve as a colorful backdrop for hundreds of other plant species throughout the gardens. Most are well-adapted to water-saving landscapes.

Accent and flowering plants are numerous as well. You'll see collections of sculptural agave, colorful aloes and succulents, plus exotic, often bizarre cacti native to each of the Southwest deserts.

Animals, Birds and Reptiles

Some of the world's most rare and exotic animals can be seen along the paths that flow naturally through the grounds. Many animals are displayed in a naturalistic setting. For example, you'll see majestic bighorn sheep on mountain slopes, Arabian oryx, Grevy's zebras and the fennel—the world's smallest fox. You'll also be treated to howling coyotes and the chatter of dozens of colorful African birds.

Home Garden Landscape Demonstration Gardens

Various gardens provide practical, take-home ideas on how to use dry climate plants in home gardens, including ways to create raised beds, mounds and slopes. Here you will find plantings of summer- and fall-blooming Texas ranger, *Leucophyllum frutescens*, and purple-flowering *Dalea pulchra;* both are introductions from Texas. A hedge has been created from a planting of jojoba, *Simmondsia chinensis*, a native of the Southwest. From Australia, you'll see the winter-flowering *Cassia* species. A raised planter near the picnic facilities features the bright colors of red salvia, *Salvia greggii*, a favorite of hummingbirds. White salvia rings the planter base. Both are native to Texas and Mexico.

Educational Programs

During the months of the year when The Living Desert is open— September 1 to June 15—children and adult programs are available. The

The smoke tree, Dalea spinosa, is one plant that truly represents the spirit of the Living Desert.

garden's philosophy is to bring the desert experience into the classroom, as well as through field excursions. Docents are also on hand on the grounds to volunteer their experience during walking group tours. Visitors will learn, for example, the many ways the deserts of the Southwest differ.

Plant Sales

Each spring, volunteers and garden staff set up a two-day plant sale. Thousands of native desert and introduced plants are on display for purchase. Home gardeners have an opportunity to purchase plants for their home gardens that are not always available commercially. In addition, the on-site plant nursery has a wide assortment of native and introduced plants ready for sale during regular garden hours.

Future plans for the Living Desert include the development of a 26,000-square-foot "Eagle Canyon," which will accommodate and care for large animals and birds such as mountain lions, bobcats, golden eagles and endangered Mexican wolves.

Honey mesquite, Prosopis glandulosa, adds a bright splash of greenery and shade to this walkway.

Desert Water Agency Demonstration Gardens

Since the early 1970s, the Desert Water Agency's conservation programs have focused on improving water efficiency in the landscape. Located three miles east of downtown Palm Springs, three demonstration and evaluation gardens are available for touring. Windy conditions, especially during spring, are a major problem. Sand dunes must be stabilized before the ground covers, shrubs and trees can become established. High-stress growing conditions, particularly hot summer temperatures, increase moisture requirements. Drip-irrigation provides deep watering for roots, helping solve this problem.

Garden One—This 1½-acre garden was established in 1981, and is designed to evaluate plant performance and water needs for the Coachella Valley region. (A key to plant names is available by writing the Desert Water Agency or pick up at the office.)

Garden Two—This 1-½-acre demonstration garden, established in 1985, is planted with water-efficient plants. The portion adjoining the building is planted with a wide variety of ground covers, shrubs and trees indigenous to many dry climate areas.

Garden Three—The 12-acre water reclamation facility was established in 1988. This facility is not open to the public, but guided tours can be arranged by contacting the DWA. Approximately 150 species are arranged in a landscape setting around the water-treatment facilities. A 3-acre parcel is used to test plant response to nutrient-rich, reclaimed water. This project, as well as Garden One, was developed in cooperation with the U.S. Soil Conservation Service.

Desert Water Agency
1200 Gene Autry Trail, South
Palm Springs, California 92264
(619) 323-4971

From I-10, exit at Gene Autry Trail. Travel south approximately 6 miles to Agency office and garden. From State Highway 111, turn north on Gene Autry Trail and travel approximately one mile to entrance.

Test gardens that provide valuable information on plant performance, in conjunction with water-efficient irrigation systems.

Open daily 8 to 5 p.m.

No entrance fee.

No picnic facilities.

Wheelchair access.

Annual events include plant sales held during spring months jointly with Desert Water Agency and the Coachella Valley Resource Conservation District. Call for date by February.

The Desert Water Agency test garden features mesquite, fountaingrass and orange-flowering bird of paradise.

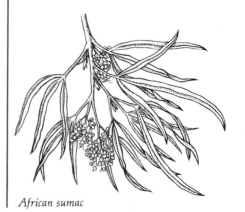

African sumac

Moorten Botanical Garden

Moorten Botanical Garden
1702 South Palm Canyon Drive
Palm Springs, California 92264
(619) 327-6555

From I-10 take Highway 111 through Palm Springs. Continue on South Palm Canyon Drive two blocks past where State Highway 111 turns east. Curbside parking.

Open Monday-Saturday 9 to 4 p.m.; Sundays 10 to 4:30 p.m.

For over 50 years, Moorten Botanical Garden has been a vital part of the Palm Springs scene with over 3,000 labeled varieties in the garden and nursery.

Nominal entrance fee.

Call for information on group tours, wedding facilities and meeting reservations.

Picnic facilities available.

Wheelchair access.

Annual events include: *Spring Wildflower Display*, March and April; *Founders Day*, Fall; *Fall Festival*, November.

Cereus cacti at Moorten Botanical Garden.

California fan palm

This is a garden that has withstood the test of time. Since 1938, the 2½ acres have been a delight to Palm Spring's residents and visitors, introducing them to the landscape of the Coachella Valley. The late "Cactus Slim" Moorten and his wife Patricia built the garden to mirror their love for the desert, with plant collections from around the world.

As you enter the garden off South Palm Canyon Drive, the historic Spanish-style residence is a focal point, set among an oasis of tall California fan palms, *Washingtonia filifera*. Immense boulders and a gentle waterfall provide a backdrop for exotic cacti and succulents.

Deserts of the Southwest

The major deserts of the Southwest are represented at the gardens. North of the entrance, trails crisscross from one desert region to another. Mature palo verde, ironwood and mesquite trees create a cool, shady overhead canopy. Plants help identify the desert regions: The high-elevation *Mojave Desert* is identified by the distinctive Joshua tree, *Yucca brevifolia*. In the middle-elevation *Sonoran Desert*, saguaro, *Carnegiea gigantea*, predominates, along with the yellow-flowering palo verde, *Cercidium floridum*. From the *Baja California Desert*, the tall, spired, boojum tree, *Idria columnaris*, among ocotillo, creates a literal "out-of-this-world" feeling. From the *Chihuahuan Desert* of Texas, tall, surreal yucca, *Yucca rostrata*, blend with silvery-leaved Texas ranger, *Leucophyllum frutescens*.

The remainder of Moorten Gardens consists of collections of barrel cacti, yuccas, prickly pear and agaves—creating a lush undergrowth of fascinating color and plant forms. Many Australian natives are currently being introduced into the gardens.

Additional Gardens of California

BEVERLY HILLS

Virginia Robinson Estate
Beverly Hills, California
(213) 276-5367

The garden's location in Beverly Hills nestles among other private estates. Visitation with guided tours are directed only through the reservation office. Address and directions to the garden are provided at time of reservations.

In 1977, Virginia Robinson Gardens became a part of the Los Angeles County Department of Arboreta and Botanic Gardens that includes South Coast Botanic Gardens, Descanso Gardens and the Los Angeles State and County Arboretum, all described in this book.

Tours given twice daily: 10 a.m. and 1 p.m., Tuesday-Thursday and once on Friday at 10 a.m. Tour time is approximately 2 hours.

Wheelchair access to grounds, although numerous steps to terraces and garden areas preclude wheelchair movement in some areas.

The Virginia Robinson home and hillside landscape represents a dramatic period in the early development of Southern California. The influence of European and Mediterranean architecture and gardens provide a unique opportunity to review and learn from the past. Expansive lawns bordered with brick walks surround the 6,000 square-foot, Beaux-Art style house and Italianate guest house. These walks lead to terraces, patios, gravity-fed ponds and fountains around the home. Numerous collections of plants and colorful seasonal displays are incorporated into formal and natural settings.

The gardens have some special features. Within the 2-acre palm grove is a dense planting of king palms, *Archontophoenix cunninghamiana*, said to be the largest stand outside of its native Australia. Other palm species, such as *Chamaedorea*, has made this grove into a near-replica of a rain forest. In tree-shaded areas, camellias and azaleas bloom late winter and spring. Rose gardens are most colorful in late spring and fall. Near the large central courtyard to the guest house, colorful English flower borders embrace the trunks of Italian cypress trees.

BORREGO SPRINGS

Anza-Borrego Desert State Park
P.O. Box 299
Borrego Springs, California 92004
(619) 767-4684 for recorded information
(619) 767-5311
(800) 444-7275 for camping reservations

From San Diego take I-8 east to Hwy 67 at El Cajon, 78 through Ramona, 79 at San Ysabel. Go beyond 76 to S2 and S22 through Ranchita to Visitors Center west of Borrego Springs.

Open daily 9 to 5 p.m. October through May; 10 to 3 p.m. Saturdays, Sundays, July 4th and Labor Day, June through September.

No entrance fee to Visitors Center. Audio/visual presentation on the hour and half-hour.

For information on wildflower season (Feb-early April) and locations write to:
Wildflowers
P.O. Box 299
Borrego Springs, CA 92004

Picnic facilities: Ask at Visitors Center.

State park rules and regulations for vehicles, campfires, firearms and dogs. Some camping areas require fees. Reservations are suggested October through May.

The unique desert environment, geological development and plant diversity within the boundaries of the *Anza-Borrego Desert State Park* is worthy of being included in a discussion of grand scale beautiful gardens. Anza-Borrego's 600,000 acres make it the largest state park in the continental United States. In this park, one feels the grandeur of the Colorado Desert terrain and native plants and animals.

Park tours are measured in the miles to be driven, trails to be hiked. During the winter season from October to May, over one million visitors explore the pine-clad, 6,000 foot mountains, travel up palm-studded canyons with California fan palms, *Washingtonia filifera*, and camp among eroded wrinkled badlands.

Recreational activities include unlimited hiking and camping, horseback riding, mountain biking (on dirt roads only—not on any hiking trails) and self-guided walking trails. Cacti, wildflower and native plant gardens with interpretive exhibits help inform you during visits. These are well planned and located around the earth-covered Visitor's Center.

Rogers Gardens
2301 San Joaquin Hills Road
Corona del Mar, California 92625
(714) 640-5800

Located near Newport Beach about 1 mile east of the Pacific Coast Highway. Take Highway 55 or the 405 Freeway to MacArthur Blvd. Head west to the address at the corner of MacArthur Blvd. and San Joaquin Road.

A retail nursery with numerous demonstration gardens, covering 7-½ acres. A place to browse for ideas or to purchase a wide selection of plants.

Open 9 to 5 p.m. January 1 to April 1; open 9 to 6 p.m. April 2 to October 15; open 9 to 9 p.m. October 16 to December 31.

No entrance fee.

Large tour groups require reservations.

Gift shop and bookstore hours: as the garden.

No restaurant or picnic facilities.

Wheelchair access.

Special annual events include: *Orchid Festival*, end of February to mid-March; *Easter Fantasy*, end of February until Easter Day; *Rose Festival*, month of April; *Fuchsia Festival*, month of May; *Christmas Fantasy*, October 15 to January 1.

This is a unique garden that artfully combines a retail nursery with extensive, colorful, demonstration gardens and landscaping ideas. The premise has appeared to have been successful for founder Gavin Herbert, who opened Rogers Gardens in 1974: over 30,000 visit the 7½-acre facility each month. Some take advantage of the full-service nursery that features a wide selection of plants and garden products; others cruise the grounds to pick up on the latest ways to use and combine plants. A traditional perennial garden, rose garden and several expansive plots of flowering ground covers, set out, side-by-side, for comparison purposes, are highlights. They even have the original gazebo found on Disneyland's Main Street, surrounded by masses of yellow daylilies, *Hemerocallis*.

A specialty of Rogers is their mossed hanging baskets, which are planted with flowering annuals. Rogers personnel offer classes on how to create the baskets, as well as instruction on other subjects. Mini-classes, as they are called, can be taken on plant care, plant selection and seasonal requirements of plants. In addition, Rogers is the home site of the PBS television show, Victory Garden West.

LANCASTER

Antelope Valley
California Poppy Reserve
15101 West Lancaster Road
Lancaster, California 93536
(805) 942-0662 (office)
(805) 724-1180 (reserve)
Wildflower Hotline:
(805) 948-1322.

Located approximately 13 miles west of the Antelope Valley Freeway (14) on Lancaster Road—an extension of Avenue I.

A 1,700-acre reserve of California poppies and many other wildflowers.

Open 9 to 4 p.m., mid-March to mid-May. Call the California Department of Parks and Recreation early in the spring months to receive prognosis of flowering season, which is dependent in part on fall and winter rainfall.

Visitor Center offers video presentation; tours on spring weekends.

Entrance fee. During off-season, Visitor Center is closed: No fee during these months.

Gift shop open seasonally as per the Visitor Center hours.

Wheelchair access to Interpretive Center, picnic area and view point; balance of trails are a walking tour.

Picnic and restroom facilities.

Only a few areas remain in California where California poppies flourish year after year. One location is in the Antelope Valley, west of Lancaster and Palmdale. The Antelope Valley California Poppy Reserve covers 1,700 acres at the 3,000-foot elevation at Antelope Buttes. This area has been established to perpetuate California poppy preservation for future generations.

The efforts of the Wildflower Preservation Committee, the California State Parks Foundation and many dedicated individuals brought about the dedication of the site in 1976. The Jane S. Pinheiro Interpretive Center was created five years later. It contains her watercolors of wildflower scenes and exhibits of California's wildflowers (12 species grow seasonally in the reserve). Over seven miles of trails loop through the Antelope Buttes.

California poppy and other annuals have evolved to bloom according to a defined sequence of sun and rain. They require a rainy, late-fall and winter, followed by a period without moisture. Until a warming trend occurs in early spring, plants lie dormant, then burst out in waves of golden blooms as the spring season develops.

LOMPOC

**La Purisima Mission
State Historical Park
2295 Purisima Road
Lompoc, California 93436
(805) 733-3713**

———

Located 4 miles east of the center of Lompoc.

———

La Purisima Mission was founded in 1787. Reconstruction of the mission began in 1934 under direction of the California Dept. of Parks and Recreation and the National Park Service. The mission was dedicated as a state historic park December 7, 1941.

———

Open October-May: 9 to 5 p.m.; June-September: 9 to 6 p.m. Closed Thanksgiving Day, Christmas Day and New Year's Day.

———

Entrance fee required.

———

Guided tours may be arranged by phone two weeks in advance. Phone (805) 733-1303.

———

Wheelchair access.

———

Gift shop: 11 to 4 p.m.

———

Picnic facilities available.

———

Facilities available for weddings by special arrangement.

———

Special events include *La Purisima Mission Fiesta Celebration*, traditional mass. Includes craft demonstrations and entertainment, held on the third Sunday in May. Phone for information on other events.

———

Nestled in Purisima Valley, outside the city limits of Lompoc beyond the beaten path of heavy traffic, you'll find the Mission La Purisima Concepcion of California as it was two centuries ago. The mission was carefully restored after the earthquake of 1812 at a location a few miles from the original site. It now presents an excursion back into the days of life during the Franciscan period.

The current garden is one acre and contains about 80 plant species, a blend of California native plants used by the Chumash Indians before the Spanish arrived, and introduced new plants.

The three gardens that comprised the original planting were replanted in the 1930s by E.D. Rowe, a nurseryman from Santa Barbara. Mr. Rowe established a garden that contained both the native plants used by Chumash Indians, and plants introduced by ther Spanish. Many plants were collected from original plantings at other missions of California.

Roses and lilies are in bloom in late May into June. In addition, the rose of Castile continues to grow and flower in abundance.

LONG BEACH

**Rancho Los Cerritos
Historic Site
4600 Virginia Road
Long Beach, California 90807
(213) 424-9423**

———

From the 405 Freeway, take the Long Beach Blvd. exit, travel north to San Antonio Drive. Go left one block to Virginia Road, then right through the Virginia Country Club to entrance.

———

Two acres of gardens and historical buildings with plants and trees that date back to 1850. "New" gardens include small, 19th century rose and herb gardens.

———

Open Wednesday-Sunday, 1 to 5 p.m. Closed Monday, Tuesday and all city holidays.

———

Self-guided tours Wednesday-Friday. Saturday and Sunday guided tours leave hourly. Reservations required for groups of ten or more.

———

No entrance fee.

———

Gift shop hours: 1 to 5 p.m.

———

Library with 5,000 historical volumes available for reference use on-site.

———

Picnic facilities available.

———

Wheelchair access.

———

Annual events include: *Easter Egg Hunt*, day before Easter Sunday; *Victorian Family Picnic*, Last Sunday in August; *Christmas Candlelight Tours*, 2 weeks before Christmas. Make reservations 3 to 4 weeks in advance.

———

See also Rancho Los Alamitos, page 40.

Rancho Los Cerritos was once a part of a 167,000-acre land grant awarded to Manual Nieto in two sections: in 1784 and 1790. Today, 4.7 acres remain, after a long history of land division. The Long Beach Department of Library Services' Historic Sites Section of the library maintains Rancho Los Cerritos. The site was dedicated as a National Historic Landmark on October 22, 1970.

Although the two-acre gardens were relandscaped in 1930 by landscape architect Ralph Cornell, many trees planted by John Temple in the 1850s, one of the many owners, still remain. In recent years, a 19th century rose garden and herb garden have been added to represent plant species available prior to 1865.

The ranch house is a two-story adobe, constructed in 1844 as headquarters for John Temple's 27,000-acre cattle ranch. Temple developed the grounds as an elaborate Italian garden, and planted many shrubs and trees—species that are still used today. Lemons, oranges, pomegranates, mission fig, olives, Italian cypress, peppers, eucalyptus and white locust were important plants, just as they were in the development of mission gardens along the coast of California.

Roses

A tour group at Rancho Los Cerritos historic site in Long Beach.

LOS ANGELES

Exposition Park Rose Garden
701 State Drive
Los Angeles, California

Mailing address:
The City of Los Angeles
Department of Recreation and Parks
City Hall East, Room 1350
200 North Main Street
Los Angeles, CA 90012
(213) 748-4772: Rose Garden
(213) 485-5515: Department of Recreation and Parks

Nearest cross streets are Exposition Blvd., Figueroa Street and Menlo Avenue. Garden is just east of Harbor Freeway.

Open daily March-November, 9 to 5 p.m. for self-guided tours. The garden is closed December-February, when roses are pruned, the soil refurbished and new varieties are planted.

No entrance fee.

Wheelchair access.

Facilities available for weddings and parties or other special events by prior arrangement. For information call (213) 548-7676.

The Exposition Park Rose Garden is surrounded by a civic cultural and recreation center and dense business and residential areas, making it convenient for visitors to stroll among the seven acres of roses. The 16,000 bushes including some 165 varieties are placed in rows and separated by grass walkways, creating an organized, formal effect. Because the garden location is near the coast, the cool air and moderate sunshine encourages a long bloom period, from spring through fall.

The garden has received special recognition from the All-America Rose Selections for its well-maintained appearance through the years. You can count on a great show of blooms almost any season. The patterns of massed color are interesting, yet observing the flowers close-up allows one to appreciate the differences in color, form and fragrance.

The rose garden began as long ago as 1911. During that period, the area was known as Agricultural Park, and agricultural fairs and exhibition of farm stock and produce were common.

The rose garden and gazebos make it a popular setting for wedding ceremonies, especially during the most-active flowering periods.

SAN SIMEON

Hearst San Simeon
750 Hearst Castle Road
San Simeon, California 93452
(805) 927-2020: Information
(800) 444-7275: Tour reservations

Visitors Center is located off Coast Highway 1, east of San Simeon Village. Entrance is well marked. From Los Angeles it's 6 hours (254 miles); from San Francisco 6 hours (245 miles).

Open daily, with first tour beginning at 8:20 a.m.; last tour at 3 p.m. (later tours during summer.) Make reservations well in advance.
Each tour takes about 1 hour and 45 minutes.
Closed Christmas Day, New Year's Day and Thanksgiving Day.

Entrance fee required.

Three or four separate tours are available, depending on time of year. Tour One is recommended for first-time visitors. Tour Four is best for garden-related aspects. Walking shoes are required.

Picnic facilities at Visitors Center.

Restaurant hours: 8:30 to 4:30 p.m.

Wheelchair access available on special tours. Call for information.

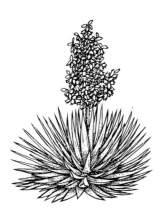

Mojave Yucca

The magnitude of the Enchanted Hill, set among the rolling coastal hills covered with coast live oaks and above sparkling ocean views, makes Hearst Castle a fascinating place. The Visitors Center at the base of the hill just off Coast Highway 1 provides the starting point for your tour, as buses wind up the hill to the Castle.

Pools, statuary, and fountains are integral parts of the gardens, as are formal hedges, roses, azaleas, citrus trees and the tall, ambient clusters of Mexican fan palms, *Washingtonia robusta*. The dramatic changes in elevation also help make the gardens special. Extensive flights of stairs lead to large courtyards that provide extended vistas to ocean views and rolling hills that stretch out and below the Castle.

The project began in 1919, with construction continuing for 28 years. Today, restoration is an on-going process.

Due to the popularity of this landmark, especially during spring and summer, plan your visit in advance—up to 8 weeks for individuals; about 12 weeks for groups. Various tours are available for various parts of the Castle. Select Tour One for an overview. Select Tour Four for a more garden-oriented tour.

TWENTY-NINE PALMS

Joshua Tree National Monument
74485 National Monument
Twenty-Nine Palms, California
92277-3597
(619) 367-7511

Located east of Yucca Valley. From Los Angeles, take I-10 east past Banning to Highway 62 to Joshua Tree west entrance. In Twenty-Nine Palms the east entrance Visitors Center holds information on travel, regulations and camping. Traveling east of Indio on I-10 (before you reach Chiricco Summit), take Cottonwood Springs Road offramp to reach the south entrance.

Visitors Centers open 8 to 4:30 p.m.

Fees apply from October-May. No fees from June-October.

Maximum length of stay in campgrounds is 30 days in summer months; 14 days during fall, winter and spring. Reservations at Black Rock and three group campgrounds can be made through Ticketron. Others are first-come, first-served.

Note: Carry in drinking water, although it is available at Blackrock and Cottonwoood campgrounds. Wood fuel must be carried into all campgrounds.

The giant, granite rock forms in Joshua Tree National Monument have been assaulted and shaped by wind, weathering and infrequent action of water to produce some of the desert's most spectacular scenery. Mountain ranges set the stage on a grand scale for many desert plants. The area also supports resident and migratory birds, small animals and reptiles. Even desert bighorn sheep live in higher elevations.

In this unique desert environment, highly specialized forms of plant life have evolved, adapting themselves to the drought conditions. Most obvious are the multi-armed Joshua tree, *Yucca brevifolia*; cane-like ocotillo, *Fouquieria splendens*; wispy smoke tree, *Dalea spinosa*; yellow-flowering palo verde, *Cercidium floridum*; and pinyon pine, *Pinus edulis*.

At Twenty-Nine Palms Visitors Center, tall, heavy-trunked California fan palms, *Washingtonia filifera*, create a cool, inviting oasis. Many kinds of cacti, including prickly pear, barrel cactus, cholla and Mojave yucca, *Yucca schidigera*, can be seen. In the spring, if rainfall has been adequate during the fall and winter, wildflowers often put on a show of color.

Selected growers and retail nurseries offer visitors the opportunity to visit their gardens and growing grounds. The following lists several located in Southern California. Often, business hours and the season the gardens are in operation are limited. For these reasons, many gardens can be visited by appointment only. Be sure to make plans to visit in advance, and call ahead to verify the garden is open for visitation.

CARLSBAD

Bird Rock Tropicals
6523 El Camino Real
Carlsbad, CA 92009
(619) 483-9393
Specialists in *tillandsias*. Garden open Saturdays, 10 to 4 p.m. Call before visiting.

CARPENTERIA

Abbey Gardens
4620 Carpinteria Ave.
Carpinteria, CA 90313
(805) 684-5112
Cacti and succulent garden. Open Tuesday-Sunday, 9 to 5 p.m., all year.

CHULA VISTA

Pacific Tree Farms
4301 Lynwood Drive
Chula Vista, CA 92010
(619) 422-2400
Wide selection of container-grown fruits, flowering trees and conifers. Specializes in new and rare plants, plus books and tools. Garden open Wednesday-Monday.

Tiny Petals Nursery
489 Minot Avenue
Chula Vista, CA 92010
(619) 422-0385
Wide selection of miniature roses, including climbers, trailers and micro-mini varieties. Open daily 9 to 5 p.m. Closed Tuesday.

CLAREMONT

Wildwood Nursery
3975 Emerald Avenue
PO Box 1334
La Verne, CA 91750
Specialists in plants for low-water landscapes, including natives from California, Australia and South Africa. A small but special nursery, planted on a former fruit orchard. Open 9 to 5 p.m. Monday-Saturday.

ENCINITAS

Weber Nursery
237 Seeman Drive
Encinitas, CA 92024
(619) 436-2194
Specialists in California natives, as well as a few plants from northern Baja California. Plants in their demonstration garden are mature or near-mature size. In spring, more than 20 species of *Ceanothus* are in bloom. Open Monday-Saturday 8 to 4 p.m. Closed July 4th, Christmas Eve and Christmas Day, New Year's Eve and New Year's Day.

Weidner's
Begonia Gardens
695 Normandy Road
Encinitas, CA 92024
(619) 436-2194
Begonias are their speciality. Select plants from 3 acres planted with 25,000 tuberous begonias, either potted or "dig your own." Petting farm and picnic tables are also on site. Open daily April-September 15, and November 1 to December 22, 9:30 to 5 p.m.

ESCONDIDO

Tropic World Nursery
26437 North Centre City
Parkway
Escondido, CA 92026
(619) 746-6108
More than 3,000 varieties of cacti and succulents, 635 roses, 500 kinds of fruits, berries and nuts, and more. Unique cactus garden features plants massed like annuals or perennials, placed in geometric designs. Certain sections of grounds available for tour by appointment only. Garden open daily 9 to 5 p.m. Closed Thanksgiving, Christmas and New Year's Day.

FOUNTAIN VALLEY

Roger & Shirley Meyer
16531 Mount Shelly Cl.
Fountain Valley, CA 92708
(714) 839-0796
Two locations available for tours. One is a home in coastal Orange County, where over three dozen exotic fruit trees, shrubs and vines are grown, demonstrating how exotic fruiting plants such as kiwi and jujube can thrive in this region. Second facility is a 10-acre farm in northern San Diego County, where several kinds of rare fruit are grown and packed. Call for appointment to visit either garden.

GARDENA

Rainforest Flora
1927 W. Rosecrans Ave.
Gardena, CA 90249
(213) 515-5200
A selection of tropical plants, including bromeliads, *tillandsias*, cycads and others. Garden open Monday-Saturday; closed major holidays. Call for hours.

JAMUL

La Quinta Botanica
Kennerson's Jamul
Nursery
14914 Lyons Valley Road
Jamul, CA 92035
(619) 669-0079
A naturalistic collection of a wide range of plants—cacti and succulents, tropicals and drought-tolerant species. Garden open 9 to 5 p.m. daily by appointment only.

NORTH HOLLYWOOD

Spencer M. Howard Orchid Imports
11802 Huston Street
North Hollywood, CA 91607
(818) 762-8275
One of the few species-only orchid collections. Garden features 12,000 species orchids from all over the world, including more than 1,000 different kinds. Open 9 to 5 p.m. daily by appointment.

Guy Wrinkle Exotic Plants
11610 Addison Street
North Hollywood, CA 91601
(818) 766-4820
Plants for collectors—over 70 species of cycads, rare succulents, haworthia, euphorbia, palms, bromeliads, orchids and ferns. Garden is open year-around, by appointment only. Call for hours.

NORTHRIDGE

Singer's Growing Things
17806 Plummer Street
Northridge, CA 91325
(818) 993-1903
Specialists in unusual succulents and plants suited to bonsai, as well as water-conserving plants. Garden and greenhouse open Friday and Saturday 9 to 5 p.m.

RAMONA

Ramona Gardens
2178 El Paso Street
Ramona, CA 92065
(619) 789-6099
Daylilies and iris are specialties; hundreds of varieties of each are on site; some daffodils as well. Open during April-June by appointment.

REDLANDS

Rhapis Palm Growers
31350 Alta Vista Drive
Redlands, CA 92373
(714)794-3823
Specialists in *Rhapis,* a type of exotic palm. More than 50 named variegated and all-green cultivars from Japan. Garden is open daily, by appointment only.

SAN GABRIEL

Stewart Orchids
1212 E. Las Tunas Blvd.
San Gabriel, 91778
(213) 283-4590 and
(818) 285-7195
Wide selection of orchids in bloom—hundreds at one time can be seen in a large display greenhouse. Open Monday-Friday 8 to 5 p.m.; Saturday 10 to 5 p.m.; Sunday 12 to 5 p.m. Closed major holidays.

SANTA BARBARA

Santa Barbara Orchid Estate
1250 Orchid Drive
Santa Barbara, CA 93111
(800) 553-3387
As the name indicates, orchids, particularly cymbidium and species orchids, are on display. Over 2,000 kinds, both common and rare. Hours are 8 to 5:30 Monday-Saturday; 10 to 4 p.m. Sunday.

Santa Barbara Water Gardens
160 E. Mountain Drive
Santa Barbara, CA 93108
(805) 969-5129
Water garden plants, including water lilies and bog plants. Garden is open Wednesday and Saturday 9 to 4 p.m.

SANTA MARGARITA

Las Pilitas Nursery
Star Route Box 23 X
Las Pilitas Road
Santa Margarita, CA 93453
(805) 438-5992
A wide selection of California native plants for low-water landscapes. Garden is open every Saturday through the year, or call for appointment.

SOLVANG

Zaca Vista Nursery
1190 Alamo Pintado Rd.
Solvang, CA 93463
(805) 688-2585
Specialists in African violet plants. Garden is open 9 to 5 p.m. Wednesday-Sunday year-around. Closed Thanksgiving Day and December 24 to first Wednesday of new year.

UPLAND

Van Ness Water Gardens
2460 N. Euclid
Upland, CA 91786
(714) 982-2425
Everything imaginable for water gardens—from a wide selection of plants to fish. Fresh-water, ecological balance specialist. Garden is open 9 to 5 p.m. Tuesday-Saturday.

VISTA

Exotica Rare Fruit Nursery
2508-B E. Vista Way
Vista, CA 92083
(619) 724-9093
Tropical plants—fruits, flowering trees and nuts—are specialities of this garden. Open all year, every day. Call for hours.

Grisby Cactus Gardens
2354 Bella Vista Drive
Vista, CA 92084
(619) 727-1323
Rare cacti and succulents for the collector, and also features large landscape specimens several feet tall. A number of greenhouses open to tours. Garden is open 8 to 4 p.m. Tuesday-Saturday, April-June.

Rainbow Gardens Nursery & Bookshop
1444 E. Taylor Street
Vista, CA 92084
(619) 758-4290
A pleasing combination of garden and bookstore. Plant specialties include selected cacti—epiphyllum, hoyas and haworthia. Garden is open 9:30 to 4 p.m. Tuesday-Saturday, by appointment only. Closed Sundays and Mondays.

WHITTIER

Greenwood Daylily Gardens
4905 Pioneer Blvd. #10
Whittier, CA 90601
(213) 699-8144
Daylily (*Hemerocallis*) garden, open only the first and third Saturday during March-November. Call for hours.

YORBA LINDA

Pixie Treasures Miniature Rose Nursery
4121 Prospect Avenue
Yorba Linda, CA 92686
(714) 993-6780
Miniature roses are this garden's offerings. More than 125 varieties and 20,000 plants on display. Open 9 to 5 p.m. Monday-Saturday.

NEVADA
GARDENS

*A desert scene at Ethel M
Botanic Garden.*

Ethel M Botanic Garden

On your next trip to Las Vegas, enjoy a special horticultural *and* taste treat: Visit the Ethel M Chocolate Factory and Botanic Garden. The 2½-acre garden just east of Las Vegas is a marvelous collection of native and introduced dry-climate plants. Over 300 plant species are distributed throughout well-planned groupings, designated by darkly colored volcanic and sandstone rock formations. Knee-high planting mounds bring cacti, succulents, perennials and shrubs into focus for close-up viewing.

Visitors can take leisurely strolls along the well-defined paths of the gardens. Most plants are clearly labeled with both common and botanic names. Moving through the garden, many flowering trees, shrubs and cacti lead your eye from one part of the garden to another. The garden's composition—a subtle blending of colors, textures and forms—gives it a pleasing well-established order.

The "backbone" plants of Ethel M—those that are an integral part of southern Nevada gardens— include the giant saguaro, *Carnegiea gigantea*, century plants, agaves, aloes, mesquite, palo verde, acacia, eucalyptus, bird of paradise, *Caesalpinia gilliesii*, teddy bear cholla, *Opuntia bigelovii*, Joshua tree, *Yucca brevifolia*, fairy duster, *Calliandra eriophylla*, and desert cassia, *Cassia eremophila*.

The climate of the high-elevation Mohave Desert in the Las Vegas area is often harsh, with strong winds, cold winters and hot summers. Plantings in this garden demonstrate how native desert plants can thrive when given sufficient moisture through drip irrigation.

The Ethel M Botanic Garden provides a positive, aesthetic experience that will be remembered long after your visit to Las Vegas. And don't forget to sample some of their world-famous chocolates.

Ethel M Botanic Garden
2 Cactus Garden Drive
Henderson, Nevada 89014
(702) 458-8864

Located 5 miles from I-15 and Las Vegas Boulevard. Take Tropicana Boulevard east to Mountain Vista, go south to Sunset Way (adjacent to the Ethel M Chocolate Factory on Cactus Garden Drive in the Green Valley Business Park). Follow signs to garden entrance.

A beautiful, 2-acre display of colorful, well-maintained desert shrubs, trees and exotic cacti and succulents.

Open daily 8:30 to 5:30 p.m.

No entrance fee.

Gift shop, restaurant and store with chocolate samples open 8:30 to 5:30 p.m.

Bus tours from Las Vegas casinos include the garden in tours. Tours of the chocolate factory are also available daily.

Wheelchair access.

Creosote bush

Desert Demonstration Garden
3701 Alta Drive
Las Vegas, Nevada 89153
(702) 258-3205

From I-15 take the West Charleston
Blvd. offramp. Proceed west on
Charleston then right on Valley View,
and right again on
Alta Drive to entrance.

A garden dedicated to water
conservation. Founded by the
Las Vegas Kiwanis Club, and
rededicated by the Las Vegas
Water District and University
of Nevada Cooperative
Extension Service.

Open Monday-Friday 8 to 6 p.m.
Open Saturdays 8 to 12 p.m.

Closed New Year's Day, Washington's
Birthday, Memorial Day, July 4th,
Labor Day, Nevada Day (October
31st), Veteran's Day, Thanksgiving
(Thursday and Friday),
and Christmas Day.

No entrance fee.

Wheelchair access.

Picnic facilities available (but no
barbecues).

Special annual events include:
Gardening Under The Stars Lectures,
by University of Nevada Cooperative
Extension. Call for schedule. *Outreach
Program for Students,* University of
Nevada Cooperative Extension, during
summer months; *Beat the Peak
Summer Landscape Watering
Lectures,* June.

Desert Demonstration Garden

Located in the dry Mojave Desert, the Desert Demonstration Garden addresses the need for water conservation. This is a garden of ideas and plants that show how to create an attractive, low-water landscape. Over 160 plant species have been planted in the demonstration garden to date; more are planned for the future.

As you leave the Visitors Center at the start of your tour, note the courtyard with raised planters. It represents the *mini-oasis concept*—lush, green plants placed in one small location where high water use is localized, with lush plants placed up close where they'll be enjoyed. Continuing east, trial plots of ground covers feature creeping Oregon grape, *Mahonia repens,* prostrate junipers, ice plant, prostrate manzanita, myoporum, *Baccharis* 'Centennial,' and others.

Moving south past the wildflower areas into the major garden grounds, native desert willow, *Chilopsis linearis,* blue palo verde, *Cercidium floridum,* Chilean mesquite, *Prosopis chilensis,* and sweet acacia, *Acacia smallii,* create an attractive border along a natural wash.

Residential plots along the south perimeter exhibit many themes: Oriental herbs, roses for the senses, natural desert garden and fruit and vegetable garden designs. Various types of outdoor paving, including stabilized, decomposed granite walks, are additional landscape ideas. In the center of the garden overlooking the grounds is the amphitheater, with Mondale pines forming a lush background.

Yellow Cassia nemophylla and blue yucca, Yucca glauca, are striking in combination at the Desert Demonstration Garden.

University of Nevada Las Vegas Arboretum

The University of Nevada Las Vegas (UNLV) Arboretum is unique in that it includes all the grounds around university buildings. This is contrary to other institutions where 10 or 20 acres are set aside for a garden. There are no fences or gates so the park-like landscape is always open.

This on-site classroom concept was created by the Nevada State Legislature in 1985. The goal is to provide a location for students and visitors to view and study landscape plants in mature stages of growth. The grounds also provide a location for evaluating new plants for landscape use around the Las Vegas area. A self-guided tour brochure available at the Marjorie Barrick Museum of Natural History includes a map that identifies plant species.

Xeric Demonstration Garden

A 2-acre xeriscape demonstration garden is located just east of the Barrick building on the south side of the campus. It serves as an outdoor extension of the museum. Design ideas, such as special paving (including use of cobble and gravel), mounding, boulders and seating, are highlights. These features are complemented by an assortment of 1,500 water-efficient plants, representing 85 species from dry climates around the world. Plants in this garden have been selected to bloom throughout the year, creating a pleasing change of scenery for visitors.

University of Nevada, Las Vegas Arboretum
4505 Maryland Parkway
Las Vegas, Nevada 89154
(702) 739-3392

Travel two miles east of I-15 and Las Vegas Boulevard on Tropicana Ave. to Swenson Ave., turn north to Harmon Ave. Entrance to garden is off Harmon Ave. Turn south at Gym Road to parking areas located north of Barrick Museum of Natural History.

A unique, on-campus arboretum and xeriscape demonstration garden.

Open daily—does not close.

No entrance fee.

Call for information on guided tours.

Wheelchair access.

Parking in visitor areas adjacent to Barrick Museum of Natural History. Self-guided tour brochures are available at the Museum office. Hours are Monday-Friday 9 to 4:45 p.m.; Saturday 10 to 4:45 p.m.

Special events are scheduled throughout the year. Call (702) 739-3392 for dates of demonstrations and tours.

The UNLV Arboretum includes all the grounds around the campus.

Chinese pistache

NEW MEXICO
GARDENS

*Interior of the UNM Biology
Greenhouse.*

University of New Mexico Department of Biology Greenhouse

Located on the campus of the University of New Mexico, the Department of Biology greenhouse is open to tours by the public. Three adjacent greenhouses nearby are not open to the public; these are used for research and teaching purposes, and for propagation of greenhouse plants. In addition, a small patio planted with cacti, native and naturalized plants, as well as some exotics, is available outside the basement level of the Biology Building.

About 270 species of plants grow in the greenhouse. One outstanding specimen, a Norfolk Island pine, towers over 45 feet high. In addition to its dramatic appearance, it creates a cooler microclimate for the more heat-sensitive ferns and orchids. Other plants in the greenhouse include bromeliads, tropical house plants, numerous cacti, *Euphorbia* and *Aloe* species.

The greenhouse is popular with the community. Here in Albuquerque's high desert climate, school children, students and the public are frequent visitors, interested in learning more about the world of exotic plants. The extensive plant collection is well-labeled and includes the family and species name, origin and common name. Greenhouse staff also provide tips on plant care and help identify plants.

University of New Mexico Department of Biology (Greenhouse) Castetter Hall Albuquerque, New Mexico 87131 (505) 277-3411 (Biology Dept.)

Greenhouse is at the north entrance to Biology Building, which is located northwest of the corner of Yale Blvd. and Central Ave. Visitors parking at lot (parking meters) near Central Ave. and Cornell Drive, east of Biology Building.

Open Monday-Friday 8:30 to 4:30 p.m. Closed university holidays: Christmas week to New Year's Day, Memorial Day, July 4th, Labor Day and Thanksgiving (Thursday and Friday).

No entrance fee.

Wheelchair access.

No gift shop, but the UNM bookstore is just northeast of greenhouse.

Soaptree yucca

**Living Desert
Zoological & Botanical
State Park**
1504 Skyline Drive
Carlsbad, New Mexico 88220
(505) 887-5516

Mailing address: PO Box 100
Carlsbad, NM 88220

Located a short distance northwest of
the City of Carlsbad. From Carlsbad,
take Highway 285 north to signs. Turn
west and follow road to
Visitors Center.

1,107 acres, of which 35 acres
are developed. Includes exhibits
on both plants and animals of the
Chihuahuan Desert.

Open 9 to 5 p.m. from Labor Day-May
14. (Last tour 4 p.m.) Open 8 to 8
p.m. May 15-Labor Day
(Last tour 7 p.m.)

Entrance fee required. Discounts apply
to adult groups of 20 or more, and to
educational groups.

No restaurant facilities, but
refreshments are available.

Gift shop at Visitors Center.
Hours as the park.

Picnic facilities.

Wheelchair access.

Highlights of annual events include:
Mescalero Apache Mescal Roast Pit,
(baking of native agaves), mid-May;
Spring Plant Sale, April.
The Mescalero Apache Mescal Roast
is part of the cultural history of the
Pecos River Valley and Guadalupe
Mountains. Mescal plants, more
commonly known as spike-leafed
agave, *Agave neomexicana,* are
prepared and roasted in rock-lined pits
during the four-day ceremony.

*Javelina run free in their arroyo
home at the Living Desert. In
the foreground is Opuntia cacti.*

Living Desert Zoological & Botanical State Park

This park presents a slice of the plant and animal life of the Chihuahuan Desert, which extends from northern Mexico into southwest Texas and southeast New Mexico. Paths and trails covering about 1½ miles lead visitors through the varied terrain found in the Chihuahuan—through sand dunes, desert uplands and *gypsum* hills.

Natural Animal Exhibits

The exhibits are designed to present a naturalized blend of animal and plant life, so one can get a true feel of how each exists in the wild. For example, an *arroyo,* a desert wash, is home to a family of javelinas. (See photo below.) Prairie dogs dart and bustle in their own village. The Tunnel Exhibit goes underground so you can view shy nocturnal animals such as kangaroo rats and ringtails. Expansive exhibits are also devoted to the rare Mexican wolf and black bear. All told, more than 50 species of animals, most which are native to the Chihuahuan Desert, can be seen on the 35 acres of developed grounds.

Plants of the Living Desert

Each desert region has its own plants that are part of its identity; The Chihuahuan is no exception. Here you'll find yucca, *Yucca* species, sotol, *Dasylirion* species, mescal, *Agave* species, prickly pear, *Opuntia* species, and *Acacia,* to name a few. In addition to Chihuahuan natives, there are extensive, indoor exhibits of cacti and succulents from around the world. Other highlights include exhibits on minerals, archeology, insects, mammals, reptiles, history and natural resources.

ALBUQUERQUE

Albuquerque Garden Center
10120 Lomas Boulevard NE
Albuquerque, New Mexico 87190
(505) 296-6020
Mailing address: PO Box 3065
Albuquerque, NM 87190

From Highway 40, take Lomas Blvd. east past Easterday Ave. to garden entrance.

Landscaped garden grounds and meeting place of the Council of Albuquerque Garden Clubs, Inc.

Open 9:30 to 2:30 p.m. Monday-Friday. Closed major holidays and last two weeks in December.

No entrance fee.

The garden shop offers gardening supplies, books and periodicals for gardeners and floral designers. Hours as the garden.

No restaurant facilities.

Picnic facilities available.

Wheelchair access.

Facilities available for meetings, luncheons and other events by reservation.

Special annual events include: *Cacti and Succulent Show*, 3rd week in April; *Garden Fair & Plant Sale*, 4th week in April; *Rose Show*, 1st weekend in June; *Spring Flower Show*, 2nd weekend in June; *Mum Show*, 3rd weekend in October.

Mescal

The Albuquerque Garden Center is a meeting place for almost two dozen garden clubs, plant societies and other gardening-related groups in the Albuquerque area. It is a clearing house for information, offering classes and information on how to garden in the greater Albuquerque region. Short courses on horticulture, home landscaping and artistic design are available.

Additional services include a gardening Hot Line, staffed by Master Gardeners, to answer gardening-related questions. Soil testing is also available for a nominal fee. The library contains more than 1,500 volumes, the most comprehensive in the Albuquerque area.

The Garden Center Building is landscaped with a patio area, small rose garden, wildflower garden and perennial bed. A map identifying the plants on-site is available at the center.

LAS CRUCES

New Mexico State University
Botanical Garden
Department of Agronomy &
Horticulture, Box 30.
Las Cruces, New Mexico
88003-0003
(505) 646-3405

Located on University Avenue, just west of I-10 and Main. Take I-10 to University Ave., then head west to entrance.

Over two acres of gardens and demonstration plots.

Open daily from sunrise until sunset.

No entrance fee.

No restaurant or picnic facilities.

Wheelchair access.

Facilities for weddings available, including a garden gazebo.

Highlights of annual events include *Open House*, held in October.

This relatively new garden, opening in 1984, is in its formative years, yet is an integral part of the NMSU ornamental horticulture program and the community of Las Cruces. It is a place where students and home gardeners alike can research and view a number of plants, side-by-side, to evaluate their growth habits and cultural requirements for the southern New Mexico area.

Presently, there are almost 100 varieties of annuals and perennials, over 130 woody plant species and 22 turfgrass plots. The garden is currently being expanded and the number of plants being grown will soon increase, as well as the overall size of the garden. Special emphasis will be given to increase the awareness and landscaping possibilities of plants native to southern New Mexico.

Commercial Gardens

BELEN

Mesa Garden
PO Box 72
Belen, NM 87002
(505) 864-3131
An extensive collection of succulent plants from the American and African semi-deserts. Visitors to the garden are welcome but strictly by advance appointment only.

New Mexico
Cactus Research
1132 E. River Road
Belen, NM 87002
(505) 864-4027
Displays of cacti and succulents—indoors in greenhouse and outdoors—including winter-hardy cacti. Rare seeds and exotics available for purchase. Call ahead to visit.

DEMING

Desert Nursery
1301 S. Copper
Deming, NM 88030
(505) 546-6264
Cacti and succulents. During March-September, garden is open six days a week. Other times of the year call ahead before visiting.

Index